Cat Miracles

INSPIRING TRUE TALES
OF REMARKABLE FELINES

Brad Steiger &
Sherry Hansen Steiger

ADAMS MEDIA CORPORATION
Avon, Massachusetts

Published by
Adams Media Corporation
57 Littlefield Street, Avon, MA 02322. U.S.A.
www.adamsmedia.com

ISBN: 1-58062-774-9

Printed in Canada.

J I H G F E D C B A

Library of Congress Cataloging-in-Publication Data
Steiger, Brad.
Cat miracles / by Brad Steiger and Sherry Hansen Steiger.
p.   cm.
ISBN 1-58062-774-9
1. Cats--United States--Anecdotes. 2. Cat owners--United
States--Anecdotes. 3. Human-animal relationships--United
States--Anecdotes. I. Steiger, Sherry Hansen. II. Title.
SF445.5 .S73   2003
636.8'00973--dc21   2002153714

This publication is designed to provide accurate and authoritative information with regard to the subject matter covered. It is sold with the understanding that the publisher is not engaged in rendering legal, accounting, or other professional advice. If legal advice or other expert assistance is required, the services of a competent professional person should be sought.
— From a *Declaration of Principles* jointly adopted by a Committee of the American Bar Association and a Committee of Publishers and Associations

Many of the designations used by manufacturers and sellers to distinguish their products are claimed as trademarks. Where those designations appear in this book and Adams Media was aware of a trademark claim, the designations have been printed with initial capital letters.

While all the events and experiences recounted in this book are true and happened to real people, some of the names, dates, and places have been changed in order to protect the privacy of certain individuals.

Illustration by Renoir/National Gallery of Art/Superstock.

This book is available at quantity discounts for bulk purchases.
For information, call 1-800-872-5627.

Dedication

It has been often said that you don't choose a cat, it chooses you. Over the years, we have been blessed by having a number of unique, graceful, elegant, colorful, and intriguing feline personalities enter our lives and choose to brighten and enrich our lives. In some cases our relationships were fleeting, but nonetheless intensely meaningful. In other instances, there was time to explore depths of consciousness and spirit that permitted us to understand more completely the beautiful awareness of the Oneness of all living things. We therefore dedicate *Cat Miracles* to our feline friends of past and present, specifically Midnight, Demon, Cleopatra, Bonaparte, Oliver, Foxy, Bart, Fred, Leonard, and Pretty Girl.

Acknowledgments

We are greatly appreciative of the time and efforts of the following individuals who contributed their wonderful stories to this book: Mark Andrews, Timothy Green Beckley, Mary Beninghoff, Clarisa Bernhardt, Dr. Bruce Goldberg, Janice Gray Kolb, Dr. Franklin Ruehl, Rev. Robert Short, Dr. Ingrid Sherman, and Jack Velayas. For all those "silent" contributors who chose to remain anonymous, thank you—you know who you are.

$\mathcal{M}$ahgy (sounds like "doggie") is my cat. I met her at a no-kill pet shelter in Chicago in 1988, six few months after I had lost my feline companion of more than twenty years, Fat Baby. I had stopped in to leave a donation in Fat Baby's memory, and I was talked into looking at their resident cats. I had sworn I'd never have another pet (famous last words) but I agreed.

The kittens were adorable, but they said that I would have to take two. It was a rule of theirs to keep the kittens from being lonely. I decided since I was there, I would go upstairs to look at the adult cats, knowing I wouldn't be tempted by a grown animal (more famous last words).

I was intrigued by a small black cat lying on top of an introduction cage. Every time I reached for it, I was greeted with hisses and growls. When I turned to look at the others, however, I felt a tap on my back. When I turned back, the cat had moved closer to me, but it still growled. This happened three times, and I finally asked the attendant about the belligerent little feline.

Her name was Mahogany, and she had been found about a year ago. She had crawled into the engine of a car, trying to get warm, and she had cuts on her hip and had obviously been abused. She was about two and a half years old when they found her and about three and a half when I met her. She'd been adopted three times but had been returned for biting, and she was now labeled "unadoptable."

There was something so appealing in her eyes that I couldn't resist. I told them I would take her. After trying to persuade me not to, two attendants got her downstairs for the vet to examine her, and it took two more to put her into the carry box for her trip home.

When I opened the box at my apartment, my new cat bolted under the bed and I didn't see her again for weeks. She ate and used her litter box, so I knew she was all right, but no coaxing could get her out from under the bed. I figured it was okay. We were two old maids living together, and she could have her space and I would have mine. At least she was safe, and it felt good to have another living being in the apartment.

I was reading one evening, almost three weeks later, when I saw her little head sticking out from under the dust ruffle. I started reading out loud. She watched me for an hour. When I got up, she retreated. This went on every night for a week or so until one evening she came out and sat by the bed about four feet from me. I continued to read and I inserted the name I had given her into the reading. I had shortened Mahogany to Mahgy, trying to get her used to the sound.

It was two months to the day I brought her home, when she startled us both by jumping onto the arm of my chair. She was only there a few seconds, but I felt we'd made a breakthrough. After that, for another month, she would sit on the chair arm and watch me when I read to her. She looked up when I said Mahgy.

One evening, instead of settling on the arm, she crawled into the space between the chair arm and my leg and stretched out full length. I was still hesitant to touch her, remembering her previous reaction, so I waited for her move. It came when she got on the bed with me one night and stretched out along my back. We had made it.

The first time I touched her, she tensed and then relaxed as I stroked her gently. After about four months from our meeting, I could finally pick her up and take her to Fat Baby's vet, Dr. Rubin, a well-known cat expert who often appears on Oprah Winfrey's show. Mahgy was so good in the car. She relaxed beside me on the seat and again in the doctor's office, and the doctor said she had obviously been cared for at one time. When he examined her mouth, however, Dr. Rubin became furious. The shelter had obviously not checked her teeth. She had gingivitis so bad that her gums were bleeding. Her teeth were broken and several were loose and ready to come out. He said it was no wonder she was a biter and belligerent. She must have been in constant pain.

Over the next six months, the doctor removed five molars and several broken teeth. He used a tranquilizer to be able to deep clean her gums and remaining teeth. The change in her personality was miraculous. The vet said she had to have been in agony, especially when she tried to eat the dry food at the shelter. I had fed her canned food, because that's what Fat Baby had eaten, and the doctor said it probably saved her life. He had no doubt she would have slowly starved on the dry food.

Within six months, Mahgy became a beautiful and loving companion. The only things she objected to were having her feet touched and claws clipped and having her belly rubbed. The doctor felt she had had frostbite at one time, and her feet were probably painful when touched. She had learned to protect her belly when she was on the street.

I retired from my job in Chicago in 1995 and moved back to Indiana, my home. Mahgy had never seen anything bigger than a rat or pigeon in Chicago, but our first night in our new home, we were greeted by about thirty ducks parading past our patio. My apartment overlooked government-protected wetlands, and wildlife teemed. I have a picture of Mahgy studying the ducks, standing still as a statue as they passed on the patio. The ducks didn't really bother her, but the small birds flitting around drove her nuts. And when the little mouse made its appearance, I thought she would go through the screen. I bought a baby gate to protect the screen door. Life was good.

Naturally I had to find a doctor for myself. When I did, he decided to try to get my diabetes under better control and put me on a new medication along with the insulin. It worked beautifully until the day Mahgy had to save my life.

I had gotten up that morning feeling rather sluggish and not well. I took the insulin, then ate a light breakfast and sat down to read the newspaper. It was cool in early November, so I covered myself with a soft throw and fell asleep.

I woke up two hours later with something pounding on my face and chest. When I opened my eyes, Mahgy was sitting on my chest, batting at my face and chest with her paws. I felt rotten. My first thought was my blood sugar. When I tested, the reading was 36 instead of the normal 120.

I immediately loaded up on orange juice and peanut butter, and an hour later I checked again. It was higher and I felt better.

When I told my doctor about it later, he said a few points lower and I could have fallen into a coma. He guessed that Mahgy had sensed that my breathing was different and reacted. For whatever reason, I was grateful and I felt that she had paid me back for any care I had given her.

But she wasn't through yet. A month later, the same thing happened. It was so sudden, I didn't realize I was falling asleep. Again, Mahgy pounded my face and body until I awoke. The reading was 46 that day.

We immediately adjusted my medication, and I have had no further problems with it. I will always be grateful for whatever made me choose Mahgy in spite of the many reasons not to. She has proven to be not only a marvelous companion, but a caring and resourceful friend as well.

I've never regretted choosing her and we are now growing old, not too gracefully, together. I'm seventy and she will soon be seventeen. We're both slowing down, but life is good. We have each other.

—MARY BENINGHOFF

Cat Miracles

*H*olly Lenz of Laguna Niguel, California, was enjoying a quiet October afternoon in the fall of 1990. She had put her two-year-old son Adam down for a nap, and she was looking forward to stretching out on a lawn chair in the backyard and catching up on some reading.

She had not gotten too far into her book when she heard what she thought was a broken sprinkler hissing. She looked about the yard, glanced toward the screen door that she had left open so she could hear Adam if he should awaken sooner than anticipated.

Suddenly, with an icy tremor of fear that shuddered through her body, Holly realized that the hissing sound was coming from inside her house. And what was even more unnerving, she could now distinguish the unmistakable buzzing sound of a rattlesnake.

Holly entered her home to encounter the horror of a four-foot-long coiled rattlesnake in the hallway. The deadly serpent had been halted at the doorway of the bedroom where little Adam was taking his nap. The only thing that held the rattlesnake at bay was their cat, Lucy.

"Clearly Lucy was protecting my little boy," Holly stated. "She was on her haunches, moving toward the snake very slowly, steadily forcing it back, away from Adam's door."

Holly called 911 and carefully moved behind Lucy toward the bedroom. She got there just in time. The toddler was up from his nap and was about to walk out into the hallway. Holly scooped up her son and carried him to safety.

Within a few minutes, the police and an animal control worker arrived at the home. The animal control officer caught the rattlesnake with a loop at the end of a pole, and a sheriff's deputy sliced off its head with a shovel blade.

Holly told reporters that the brave Lucy would thereafter receive nothing but top-grade cat food as a reward for having saved Adam from the reptilian invader.

*S*ome years ago when my mother was recovering from surgery for a broken hip, Dr. Huff, her physician, became very concerned that the healing process had not been as speedy as he had hoped. In fact, he told us that Mother might have to be returned to the hospital for closer care. Because the doctor's anxiety was soon passed along to her friends and family members, we would all gather around Mother's bed to visit and offer words of cheer. We also brought gifts of flowers, along with magazines and books to read.

Mother's cat, Menu, in my opinion, was extremely close to being "wild," as it would let no one touch it but Mother. Whenever anyone else approached her and tried to make nice, Menu always snarled as only she could, arching her back in the feline posture of attack or flight.

However, on this one special occasion, Menu, to the surprise of everyone assembled there that day, came strolling into the bedroom where my mother was recovering. Menu moved gracefully through all the visitors without a snarl and boldly hopped up onto the bedside where my mother lay.

As we all watched in wonder, Menu gingerly began to walk, slowly and carefully, across the area of my mother's body that was covered in a light cast. At this point, Menu stopped and dropped the gift of a cricket on top of Mother's pajama coat.

Once she had deposited her gift, Menu looked at Mother and gave a questioning "Meow," as if to say, "I'm bringing you a gift, too. Do you like it?" Then she swished her plush furry tail, turned to leap down from Mother's bed, ran out into the hallway, and disappeared.

Everyone, including Mother, had a hearty laugh at Menu's surprise appearance and her unexpected gift. And, strangely enough, to everyone's delight, it seemed that right after Menu had presented her gift of a cricket, Mother began very quickly to get better.

When Dr. Huff came the next evening to check in on

Mother, he was pleased to say that he noticed a great and sudden improvement in her condition. Later, when someone told him about Menu's gift of a cricket, he laughed and joked that maybe he would hire the cat as an assistant, because Mother's healing continued from that day forward.

I have often pondered whether or not Menu could possibly have observed that gifts were being brought to Mother for a purpose. What might Menu's thoughts have been? Even if she was just mimicking us humans, such an action of imitation surely indicated that Menu was able to reason that she should do likewise. I like to think that between all creatures there can be a line of communication, and somehow just maybe Menu "knew" that this was the thing to do. From her feline point of view, she brought Mother get-well wishes and the very proper gift of a cricket. Menu's special gift was talked about in our circle of family and friends for years.

—CLARISA BERNHARDT

*I*n March 1990, four-year-old Aurelie Assemat nearly died when she fell from a relative's fourth-floor apartment in Grugny, France. Although she lay in a coma for a month and her parents feared that they would lose her, Aurelie finally awoke. Sadly, though, she could no longer speak, she was nearly blind, and she was paralyzed down her left side.

Desperate to find some way to enliven their little girl's tragic condition, Georgette and Jean-Louis Assemat hit upon the idea of giving her a cat. They were delighted to see that the green-eyed tabby seemed to lift their daughter's spirits at once. Aurelie named her new friend Scrooge, and they became inseparable companions.

Georgette and Jean-Louis were greatly relieved to see that they had found an antidote to Aurelie's loneliness. She would hug the cat and cuddle it all day long.

In August 1991, the Assemat family left their home in Grugny for a vacation in the country. Scrooge seemed to like the countryside as much as his human family members did, but when it was time to go home, it suddenly appeared as though the cat had decided he liked the new environment a bit too much. The Assemats were packed and ready to return home, but Scrooge was nowhere to be seen. He had wandered off at the last minute and had apparently gotten lost.

Finally, the hour arrived when they had no choice other than to begin the trip home without their feline friend—their daughter's dearest companion. Although Aurelie had lost her voice due to her fall, she cried for her lost Scrooge throughout the entire 600-mile journey.

The effects on Aurelie of losing her cat proved to be devastating. She did not eat; she did not sleep; she lay sobbing in her bed, staring at Scrooge's empty basket. Her parents were afraid that their daughter had lost all interest in life.

As the months went by, Aurelie grew ever more listless and despondent. When they had first returned from the country, Georgette and Jean-Louis knew that their daughter had kept alive a hope that somehow Scrooge would return to her. Although they felt that such a dream of reunion was an impossibility, they said nothing to dash Aurelie's faith.

Then, on July 9, Aurelie and Georgette heard a feeble scratching at their front door. Aurelie began to move her wheelchair toward the door with an excitement that she had not experienced for nearly a year.

When Georgette opened the door, they were astonished to behold a woefully bedraggled cat. Then, for the first time since her accident, little Aurelie spoke: "Scrooge! It's my Scrooge! He's come home!"

The cat walked to the girl's side and rubbed itself against her legs. Scrooge had found his way across the 600 miles that separated him from the Assemats' home in Grugny. It had taken him nearly eleven months to do so, but he had accomplished what had seemed to Georgette and Jean-Louis to be the impossible.

Georgette remembered that she was hugging and kissing Aurelie as her daughter was hugging and kissing Scrooge.

But the determined cat had paid a terrible physical price to complete his odyssey. It was immediately apparent to the Assemats that, fighter though he was, Scrooge was on his last legs. They resolved to take him directly to a veterinarian just as soon as they had given him something to eat.

The veterinarian was advised to spare no cost to repair the rugged Scrooge, for the Assemats had received a firm realization during his absence that he was the best cure they could provide for Aurelie. The fact that she had regained her ability to speak at the very sight of her feline friend was additional proof that Scrooge was Aurelie's most effective prescription for a more complete recovery.

Scrooge's proud tail had to be amputated. The veterinarian expressed an educated guess that Scrooge had caught his tail in a trap and had yanked it free so that he might continue his journey home to Aurelie. Scrooge had gained his freedom with such a desperate act, but he had also wounded himself to the extent that his tail was riddled with gangrene. He also required an emergency hernia operation.

A few months later—after extended Scrooge therapy—Aurelie was speaking in an almost normal manner, and she had begun to walk with the aid of crutches. Without Scrooge, the Assemats dread to consider what might have become of their little girl.

Some days after he had returned to their household, Aurelie told her mother: "My love guided Scrooge home. He knew that I was crying for him, and he just followed my tears home."

Mogadon, an eighteen-month-old black-and-white cat, was probably just enjoying a pleasant spring day in April 1993 when she was apparently startled by the voice of her owner's girlfriend telling her to get back into the apartment. The problem was, Mogadon was enjoying the fresh air and sunshine from a spot on a ledge that was twenty-one stories above the streets near Leeds, Alabama.

Claire Quickmire had only the kitty's best interests at heart when she attempted to call Mogadon inside, but the cat seemed to become confused. Perhaps she was used to taking such orders only from her owner, Mike Hawksworth.

In the next few seconds, the cat had fallen from her precarious perch and was plummeting toward the sidewalk below.

Claire recalled that the ride to the sidewalk in the high-rise's elevator was the longest elevator ride of her life. She fully expected to find Mike's kitty to be nothing more than a mangled mess of blood and fur.

She was astonished to find the dazed and frightened cat being comforted in the arms of a neighbor.

According to eyewitnesses, Mogadon had streaked toward the cement, struck a bush at the side of the high-rise, then bounced onto the sidewalk. The befuddled cat picked itself up, then walked off, slightly dazed, but apparently not a great deal worse off than she was before the twenty-one-story plunge to earth.

Claire gathered Mogadon into her arms and rushed her to a veterinarian, who found no broken bones, but who located numerous cuts and bruises to testify to the miracle survival. The supercat required twenty-four stitches, which seemed a very small price to pay for having endured an impossible fall.

When Sucile's son Joe was around twelve years old, he brought home a beautiful little white kitty all covered with grease and told her that the kitten was her birthday gift. Sucile's response was intended to be very firm: "Honey, we just can't have a cat."

"But, Mom"—Joe frowned—"It's *your* birthday gift."

So they cleaned the little kitty up. Sucile wrapped it in a towel and put it on her lap. "She just curled up and seemed at home," she said. "By that time, I just couldn't give her up, so we kept the cat. Joe is twenty-three now, so you can figure out how long we had the cat."

Sucile and her children were living in South D[akota] when the cat she christened Lucky came into their [lives.] As the years went by, she began to notice that Luck[y] seemed to be very psychic or was a very good weath[er] predictor. "If it was a couple of hours before a storm [or] tornado was coming, she would run up my wall," Suc[ile] said. "I finally realized that she was warning me in advance of every storm."

Lucky liked to lie on Sucile's stomach while she read or watched television. "But when I came home after having had some surgery on my stomach, Lucky seemed to know and would not lie on my stomach, but only on my neck," Sucile said. "Sometimes I would say to her, 'If you love me, blink your eyes.' She would blink them twice, then look up at me and purr, touching me lightly with her paw."

After a few years, Sucile got another kitten as a play-mate for Lucky, and she felt that both of the cats consid-ered themselves her loyal guardians.

"The day that they gave their lives for me started out as a very quiet and peaceful summer's day," Sucile said. "I was sitting in my home in South Dakota, writing a letter, and Lucky and her playmate were sitting on the window ledge, watching me. Suddenly a big storm came up, and I heard a loud crack as lightning struck a nearby tree. I looked up to see a big red ball outside my kitchen window. Somehow the lightning bolt had created the dangerous phenomenon of ball lightning. That deadly charge of electricity appeared about to pass through the window and to come directly for me."

Sucile will always remember how her beloved cats positioned themselves so that their bodies intercepted the ball lightning. "The cats just lit up. Their entire bodies glowed, and there was a bluish white aura around them as the lightning touched them. Afterward, when the firemen came, they said they could not believe that the charge had not continued through the cats and hit me."

Sucile knows in her very core of being that her cats saved her life by intercepting the lightning's deadly charge. Miraculously, Lucky and her kitty friend had survived the terrible shock, but they had lost their hair and were badly burned.

"My veterinarian told me that he knew that I loved my kitties, but if I really loved them, I would put them to sleep," Sucile concluded her story. "I allowed him to do so to prevent my darling cats from enduring any more suffering on my behalf."

*A*fter several days in her
new home, Winnie
Wagner of Orange
Park, Florida, decided that either a cat was somehow
trapped inside the wall near her bathtub or she had lost
her mind.

Finally, in late February 1994, she had the marble
paneling removed from around her tub, and a small cat
blinked at them from inside the wall.

Winnie remembered that she felt faint when there
was no longer any doubt that there really was a cat
trapped inside the wall. She thought of all the nights that

she had heard his pitiful cries—and all the while he was just inches away from her. "Wally," as he was appropriately christened, was carefully removed from the bed that he had made of insulation, and Winnie took him to Briarcliff Animal Clinic in nearby Jacksonville.

Dr. Susan Ridinger weighed Wally in at only three and a half pounds, but stated that although dehydrated and weak, he was in amazingly good shape, "considering what he has been through."

The veterinarian theorized that Wally had lived off condensation from pipes under the tub in Winnie's bathroom, but doubted that the cat could have survived for much longer.

No one could venture any more than a guess as to how Wally had found himself trapped in the walls of the home, but educated estimates concluded that he had somehow managed to endure forty-five days sealed up in solitary confinement.

The story has a happy ending for Wally. More than 200 adoption offers for the amazing kitty flooded the clinic, and he was placed in a new home with a loving owner.

*N*ew mothers who are also cat owners have had to deal with that unsettling old wives' tale about evil felines that suck the breath out of babies as they lie sleeping in their cribs. Kandy Phillips is able to testify that her cat reversed the ancient myth by saving the life of her choking baby.

Kandy, who resides in Modesto, California, was in the kitchen doing dishes one afternoon in the fall of 1990 when her cat Oscar ran into the room and began a loud yowling. Before she could quiet the noisy cat, Oscar jumped up on the kitchen counter.

Irritated by Oscar's blatant disregard of one of the basic "no-no" rules of the house, Kandy brushed him off the counter.

"But even though he knew he would be punished, Oscar jumped right back up," Kandy said.

This time when she shoved him off the counter, Oscar nipped her on the leg. "He didn't bite me hard, but just enough to let me know that he was really serious about something. Then Oscar started making a lot of noise and began running around in circles."

Totally puzzled by Oscar's bizarre behavior, Kandy became even more baffled when he ran toward the bedroom where she had just placed her son Anthony for his nap. As far as she knew, her four-month-old baby was sleeping peacefully in his crib.

Out of a mixture of curiosity and annoyance, Kandy followed Oscar into the bedroom. Once inside, she was astonished when Oscar boldly violated another basic house "no-no" by jumping up on the changing table. Then, before she could even swat him, Oscar had bounded directly into Anthony's crib.

Perhaps dreadful images of all those old tales about cats suffocating infants momentarily flashed before Kandy's mind, but she was soon crying out in a real, not imagined, terror.

Anthony was lying on his side, his face purple, his eyes tightly shut—and he did not appear to be breathing.

It was obvious that he had spit up in his sleep and had choked on the vomit.

Trembling in horror, but acting on mother's instinct, she tapped Anthony on his back and attempted mouth-to-mouth resuscitation.

Nothing seemed to help until, in desperation, Kandy held her baby upside down and hit him hard on the back.

"That did it! A stream of vomit left his mouth, and he gasped and brought air back into his lungs. Anthony's crying was music to my ears."

Once her baby was out of danger, Kandy was able to sort out the pieces of the household drama that had nearly become a tragedy. Oscar had saved Anthony's life. If he had not summoned her to the bedroom, her four-month-old son would almost certainly have choked to death.

Something had alerted Oscar to the baby's dangerous situation. Perhaps the child's unusual gagging and choking had disturbed Oscar. Or maybe the cat simply tuned in telepathically to the infant's distress.

Whatever the motivating cause, Oscar was so upset by the baby's situation that he violated all the basic house "no-nos" in order to signal Kandy Phillips that she must check on Anthony at once.

When I was growing up we had two pets that regularly shared the house. The first was a Prussian-mix cat named Heidi. She was, more than the average cat, very aloof and rather disdainful of the rest of her family. Our other indoor pet was a toy poodle named Missy that was actually smaller than the cat. Yet she was very feisty and insisted on being at the center of whatever was going on.

In all the years that we had these two pets, I do not recall any time that they had anything to do with one another. They were not enemies, as might be expected of a cat and dog, but nor did they play together.

One day my mother, my father, and myself were in the front yard doing some gardening. Missy had come out to be part of the action. While we were working, the neighbor's dog, a pit bull mix, began to stalk Missy, approaching her from behind.

I recall seeing the dog approach as I knelt pulling weeds. I had just started to get up to get Missy, to put her in the house out of harm's way, when the pit bull, who could have killed our dog with a single bite, began its charge.

Before I could react, Heidi sprang from the bush in which she had been hiding and intercepted the attack. She scratched at that big dog like a buzz saw! Surprised, the attacker turned and ran away with Heidi chasing him and swatting his tail with every other step.

After chasing him for about 150 feet, she broke off her attack and returned home to rest in the same bush from which she had been surveying our domestic scene, as though nothing had happened. Our Missy went into the house. All was as it had been.

Both the cat and dog lived for several years, and we never saw them interact in any other way. It was impressive though to see that the cat, though acting aloof, was in fact ready to protect her family with her own life.

— JACK VELAYAS

*A*lthough Clarence Johnson of Jacksonville, Florida, was the proud and contented owner of many fine pets, he had long cherished his favorite cat, Zeus, a lordly and beautiful seal point Siamese that he had raised from a kitten.

"There is so much satisfaction to be gained from raising a magnificent animal from a tiny ball of fur to a full-grown lord of the household," Clarence stated. "I had personally nursed Zeus through a number of minor health problems, and he had rewarded me with the greatest gift that a pet can bestow upon a human being—

loyalty and affection."

Zeus had a peculiar habit of crawling up on Clarence's chest while he was still in bed in the morning and softly kneading the muscles of his stomach and chest. "It was as if he were gently waking me to fix his breakfast; and as kind of an exchange of energy, he was giving me a massage to help me get up and going."

Zeus was seven years old when one day in June of 1987 he became seriously ill. Clarence immediately took his pet to a well-equipped animal hospital. There he learned for the first time about cystitis, a condition that the veterinarian told him was a fairly common affliction of male cats.

"The doctor told me that surgery would be to no avail," Clarence sadly recalled. "Even the very best medical treatment would prolong Zeus's life only a very little longer. In compassionate but frank terms, the vet told me that if I were to put Zeus to sleep at that time, I would spare him a great deal more suffering. If I selfishly decided to hang on to Zeus until the very end, my beloved cat would die slowly and painfully."

Clarence was spun into a quandary. One always hopes for a miracle cure. Perhaps Zeus's condition might go into remission. What if he were to consent to Zeus's mercy killing and some new wonder drug were to be announced next week?

On the other hand, the veterinarian seemed so certain

that the act of prolonging Zeus's life would prolong his pain.

"I couldn't bear to be the agent responsible for Zeus's suffering," Clarence said. "I returned the next day and told the vet that I had decided that he should end Zeus's agony."

Once the decision was made, Clarence was engulfed by terrible feelings of guilt. "I knew I had done the right thing, but I kept having this image of Zeus somehow knowing that I had condemned him to death. Certainly he was in pain, but would he be able to understand the concept of mercy killing? Would he comprehend why I had condoned his death sentence?"

Even greater depression flooded Clarence's psyche when he began to blame himself for having been inattentive to Zeus's condition. "Perhaps if I had been more observant I would have spotted the signs of the illness early enough to have done something positive about the condition," he said. "Maybe if I had given him better care, Zeus wouldn't have developed the condition in the first place."

Then one night as Clarence lay reading in his bed prior to failing asleep, he was startled to feel something land on the bed beside him.

"I knew at once that it was Zeus," he said. "I clearly heard the familiar sound of his purring; and even though I could at first see nothing, I felt the gentle touch of his paws begin to knead the muscles of my stomach."

Clarence focused on his most pleasant memories of Zeus, and slowly, dimly, but unmistakably, he saw his pet staring lovingly into his eyes.

"I could feel the strong bond of love and affection that still existed between us," he said. "I knew then that Zeus understood what I had done and that he did not blame me in any way for making the decision that had ended his suffering. He seemed to be telling me that he knew that I had done the best that I could for him."

Clarence Johnson continued to gaze affectionately into Zeus's eyes until he fell into a deep sleep. "I did not awaken until the next morning, when I opened my eyes to embrace a new day with a total feeling of peace and love toward the wisdom of the universe."

*S*everal years ago, Kathy, one of my patients, reported a most unusual story involving a nine-year-old cat owned by her father, John. Kathy was forty years old when she went camping with her father, who was nearly seventy. John rented two small cabins in Wisconsin, and they both spent a weekend getting back to nature.

Kathy was divorced with no children, and her father was a widower, when this event took place. Both Kathy and John were very private people, so each of them having their own cabin was ideal. John always brought

Holmes, his nine-year-old cat, with him on trips.

On Saturday evening Kathy went to bed around midnight after chatting with John after dinner for several hours. Their cabins were about 200 feet apart. At about 1:30 A.M. Kathy felt a pressure on her chest and awoke to find Holmes sitting on her chest. In all the years she knew Holmes, he had never done that before. Kathy couldn't understand how and why Holmes was in her cabin. She tried to push Holmes off her chest, but he resisted. Finally, she moved Holmes off her chest and decided to check in on her father, and return his cat to him.

When Kathy entered John's cabin, she smelled smoke and headed to the kitchen. A small fire was present, and Kathy immediately extinguished it. When she finally knocked on her father's bedroom door, he did not answer.

Kathy rushed in and was shocked to see her father looking pale and paralyzed. Not wasting another moment, she called an ambulance.

John had had a stroke and would have died had it not been for the efforts of the guardian-angel feline known as Holmes. Either the stroke or the fire that began in the kitchen would surely have ended John's life.

—DR. BRUCE GOLDBERG

When Brad was a boy of thirteen, he enjoyed a remarkable mind linkup with a cat that so inspired him and so convinced him of the oneness of all life that he resolved to devote his life to communicating this revelation to others through his writings.

Brad first met the cat one late afternoon as he was doing chores on the Iowa farm on which he grew up. He was carrying a pail of pig feed in each hand when he heard that peculiar kind of purring chirp that some cats make when they are seeking attention. Strangely, this cat-speak was coming from over his head.

Brad paused to look upward and saw an unfamiliar, mottled black-and-orange cat perched on the edge of the roof of one of the livestock sheds. On the farm, cats are an absolute necessity to keep the grain bins from being completely overrun with mice and rats. Brad's family had many cats that lived in the barn and foraged around all the corn cribs and grain bins for rodents. Although few of the cats were truly what one could call tame, Brad was familiar with every one of them and had awarded names to a good many—but he had never before seen this strange cat that seemed to fancy sitting up high on roofs as if it could fly.

Once it had his attention, the cat made a move as if it were about to jump down at Brad. This should have caused concern in the teenaged farm boy, for his mother had a strong distrust of cats. While she never mistreated any of the cats on the farm, neither did she tolerate their invasion of her personal space. Brad had many times heard the story of how a feral cat had dropped from a tree and viciously attacked his mother and her older sister when they were little girls. So he definitely should have moved out of the way of its trajectory when the strange cat launched itself into the air and seemed deliberately to be aiming its body directly for him.

But there was something about that preparatory chirping purr that seemed friendly and nonaggressive.

Brad stood resolutely, a pail of feed still in each hand, as the cat landed expertly on one of his shoulders. Incredibly, its claws skillfully penetrated just enough of Brad's light jacket to guarantee the cat's balance on his shoulder without piercing any of the skin beneath them.

"Comfortable?" Brad asked his new friend, the flying cat. The answer came in a soft mewing response and a contented purring as the cat gently rubbed its cheek against Brad's in what could only be considered a friendly greeting.

Brad completed his chores, feeding the pigs, tossing down hay for the cattle, mixing feed for his 4-H calves, with the most unusual cat still riding on his shoulder. When he crawled over a fence to enter the pig lot, the cat remained perched jauntily on his shoulder. Even when he climbed the ladder to the second floor of the barn to break open hay bales for the cattle, the cat rode along, expertly shifting his weight to maintain its balance on Brad's shoulder.

And so it went for months. The cat waited for him on the edge of a livestock shed, and after the two friends greeted each other, it would jump down on his shoulder and ride there contentedly as Brad did his chores. Wherever Brad went on the farm, the cat was perched on his shoulder, like a sailor with his parrot.

A mischievous mixture of genes had swirled the color pattern of the cat's regular orange tiger stripes into peculiarly placed splotches." Because of the unsymmetrical splashes of orange shaded with black, the cat appeared to have a prominent nose and a little human-like face. Once a neighbor witnessed the cat riding on Brad's shoulder, and after beholding the tiny gargoyle-like face on the creature said that the cat looked like a little demon. Thinking more of the story of Socrates and the daimon on his shoulder that Grandma Dena had told him, rather than some hellish imp from down below, Brad named his friend, Demon. Grandma Dena, who was also the town librarian, had said that when Socrates spoke of his "daimon," it was more in the sense of a guardian angel or friendly spirit, and that was the way that Brad preferred to think of his feline companion, a very friendly and dependable spirit.

At first Demon appeared whenever Brad did his chores, leaping from the low roof of a nearby livestock shed onto his shoulder, quietly seeming to inspect the quality of Brad's work as he fed the cattle, picked eggs in the henhouse, or curried his pet calves; but later the cat would seem to materialize everywhere in an almost magical way.

When Brad's father decided to limit the number of chickens to be raised on their farm, a small chicken house in very good condition was left vacant. The building was hauled to a pleasant spot in a grove of pine trees, where Brad's nine-year-old sister June was given half of the area for her playhouse, and Brad was given the other half for his art studio.

From the time that he was eight years old, Brad's greatest pleasure lay in writing stories and illustrating them; and by the time that he was a teenager, this had become his passion. Whenever he had the briefest spare moment from baling hay, unloading corn, doing the livestock chores, or working in the fields, he would dash to his studio and work at his writing and his art. He was never at his drawing board or writing desk for longer than a few minutes before Demon would jump through the open window and sit quietly on her haunches beside him. She would never play with the art gum eraser or the pencils. She would not clean herself or stretch out for a nap. She would sit there in rapt attention and watch Brad draw or write for as long as he was able to spend at his drawing board or desk.

Because of the erratic schedule of farm work during the daytime, there were never set times when Brad might rush to his drawing board for a few minutes of artistic expression before he went back to his appointed duties. It would have been impossible for the cat to establish a habit pattern conditioned by the regularity of her human friend's routine. Brad's appearances at his little studio were totally spontaneous and completely at random. And yet, without fail, Demon would appear within minutes every time Brad sat down to write or draw.

The evening's schedule was no more consistent. One day's work might demand unloading bales of hay until long after dark. A rainy day meant early quitting time and an opportunity to begin drawing or writing right after supper. Then there were the nights that Brad's father played catcher on the town's softball team, and Brad played right field on the junior team.

The fact that Demon would appear within minutes of Brad's setting himself at his drawing board—regardless of the time, day or night—convinced the teenaged boy that he and his cat companion had established a special kind of mind attunement. It seemed illogical to think that the cat hovered ever near the converted chicken house, just waiting for the moment when Brad would enter it. A much simpler solution in Brad's evaluation of the mystery was that Demon, wherever she might be, just *knew* when her human friend arrived at his desk, and set out to meet him as soon as she received the telepathic signal.

On several nights, tired as he was from a long day's work in the field, Brad decided to test the mental linkup that he was convinced existed between Demon and himself. Even at these decidedly off hours, Brad, often accompanied by his little sister, would scarcely be in the door of the small house before Demon appeared at the windows, tapping her paw for admittance if they should happen to be closed against rain or mosquitoes.

Brad began to think of Demon as his Muse. If those

ancient Greek writers and artists had friendly spirits who inspired their works, why couldn't an Iowa farm boy whose goal was to be an artist and a writer have a special Muse in modern times? And why couldn't that Muse take the form of a most unusual and very special cat?

In retrospect, it was probably significant for the boy's psychological development that the cat wanted so much to be with him. A cat, traditionally noted for its aloofness and independence, actually preferred his company to any other's. And interestingly, the cat never appeared to want anything from him. Demon never begged for food or any kind of special attention. It seemed enough just to be with the boy—and to have an occasional shoulder to ride upon.

One day, as it often happens in life when things seem just too good to be true, Demon no longer jumped down on Brad's shoulder or appeared magically through the window of his little art studio. For the first time ever in their yearlong relationship, Brad called for her. Then, when there was no response, he searched the entire farm from corn bin to south forty, seeking his companion, his inspiration, his furry, four-legged Muse.

But no trace of her was to be found. Just as she had simply materialized that afternoon with her unique mewing chirp on the roof of a livestock shed and sig-naled her arrival in his life, Demon had now disappeared without a good-bye meow or purr.

Although Brad greatly felt her loss and felt for a time

an almost bitter sense of having been deserted by her, he came to realize the wonderful lessons that Demon had taught him, lessons which he has sought to apply throughout his life: Any life form treated with kindness, love, and respect will return that positive energy many times over—and that there is a beautiful Oneness to all of life. And whoever or whatever Demon was—a uniquely remarkable cat or a Muse in disguise come to inspire a young farm boy to seek his dream—Brad will never forget her and the magical moments that they shared.

*I*n the summer of 1993, drivers for Mollerup's Van and Storage Company in Roy, Utah, were puzzled by strange noises coming from some furniture pieces they were storing for a military family that was in the process of relocating. Once they had pinpointed the peculiar sounds as issuing from a dresser, they opened a drawer to discover a female striped cat that resembled "a bag of bones."

Terri Anglen, of Mollerup's, declared the cat's survival an "absolute miracle." Her owners had been transferred to Hill Air Force Base in Utah from their present home in Michigan, and it was apparent that the cat had been trapped in that dresser drawer for two weeks without any food or water.

When the astonished owners of the cat were contacted, they said that they had searched high and low for their feline friend before their vanload of furniture had left Michigan for Utah. What had happened to their cat had been a complete mystery to them. Terri Anglen said that the owners asked the cat's rescuers to be certain that she was fed and watered and loved until someone arrived to take her home.

"*G*inny, you old silly! What have you got into now?" Mrs. Winifred Mansell of Keston, England, demanded of her pet cat, a large orange tabby. "You're limping, poor thing. Let's come to mother to have a look, eh?"

Mrs. Mansell had been taking full advantage of the lovely summer weather that early evening in 1955 to work in her garden. Ginny had been taking full advantage of her mistress's all-consuming concern for her flowers and plants and had gone wandering off in the nearby fields. "Oh, darling," Mrs. Mansell exclaimed, the welfare of her beloved feline once again uppermost in her consciousness and the weeding of the flowers temporarily set aside. "You are limping badly. You must have picked up a nasty thorn in your paw."

When the woman lifted her tabby to examine its ailing left forepaw, she immediately spotted something shiny reflecting the fading light of early evening. Gently, she probed the paw with her right forefinger, but Ginny squirmed against the pain.

"Hold still now, love," Mrs. Mansell said, tightening her grip on the tabby.

As quickly as possible, she removed two particles that she at first believed to be chunks of glass from between the pads of Ginny's left forepaw.

But then, rubbing the tiny objects between the thumb and forefinger of her gloved left hand, Mrs. Mansell made a startling discovery: "They look like diamonds!"

The next morning Winifred Mansell was astonished to hear an experienced jeweler agree with her earlier analysis of the foreign objects that she had removed from the forepaw of her limping cat.

"Diamonds, madam," the jeweler pronounced. "Diamonds of rather good quality. Worth at least £300 (about $600) each."

Somehow, somewhere, as she roamed the nearby fields, Ginny had managed to wedge two valuable diamonds between the pads of her left forepaw.

Had she dug up some miser's old unclaimed treasure trove? Had she somehow stepped on a burglar's hidden swag? Or had she just happened to have "found" diamonds that were lost during someone's stroll across the grassy fields?

No one ever determined the source of Ginny's mysterious booty, but Mrs. Mansell could proclaim that cats, as well as diamonds, are a girl's best friend."

"Meow! Meow! Meow!" That was the melodious song my mother, Florence Ruehl, heard emanating from the bedroom one morning as she was cleaning the breakfast dishes in the kitchen. This startled her because, although inveterate ailurophiles, my parents did not own a cat!

When she ventured in to investigate, she was shocked to see a plump black tabby with a thick fluffy tail atop the dresser, pawing through her jewelry. He promptly hopped off, approached her meowing, and rubbed his soft fur through her legs. Of course, she could not resist his charms and began stroking him, inducing a loud cascade of delighted purrs. She also poured a saucer of milk for him, which he quickly lapped up.

Then, the feline introduced himself to me, his furry body against mine. At the time, I was a three-year-old toddler, and while I certainly do not remember many of the events of that period, I have a vivid recollection of enjoying playing with this kitty!

Later, my mother opened the front door and he scampered off. But, she was still thoroughly perplexed as to how he had gained entry to our domicile. At the time, we had just moved into a unit in Philadelphia that was actually the second floor of a home that had been converted into an apartment by the owners after their adult son had moved out on his own. When she mentioned the incident to them, Florence was informed that the cat was their own Billy, and that he still considered the upper level his domain, even if it was occupied by another family.

From then on, Billy paid us a regular morning social call, making the rounds of the apartment as though he were a feudal lord inspecting his estate.

After several days, Florence still had not figured out how Billy was getting in. Because this was a sweltering summer, she kept the windows open, but all were screened. One morning, she decided to maintain a vigil, awaiting Billy's arrival to find out how he entered. And that was when she learned that he was truly a cat burglar. He apparently had used his claws to widen a small tear in the bathroom window screen, and, despite his girth, was able to deftly squeeze through it and into his kingdom.

While Florence initially enjoyed his intrusions, she became decidedly unhinged one day when Billy, on his favorite spot on the dresser, swatted a compact onto the floor, creating a powdery mess for her to clean up. Then and there, she decided to block his entree by taping a piece of cardboard over the screen opening. There would be no more uninvited trespassing by Billy!

But the next morning, unbeknownst to her, Billy, with the sagacity of an animal Einstein, used his claws to push away the cardboard barrier and enter once again. No one was going to deny him entry to his realm! She was asleep on the sofa when Billy suddenly pounced onto her chest, yowling loudly and persistently. She could not believe that he was back again, and could not understand what all the caterwauling was about, which was atypical for Billy. Then, he leapt onto the floor, still wailing, as though trying to sound an alarm.

She followed him as he led her to me, lying on the kitchen floor, choking to death! My mother slapped me several times on the back (the Heimlich maneuver had not yet been invented), and dislodged a chunk of chocolate chip cookie that had stuck in my throat. Billy had saved my life, and he became an instant hero with my mother. She even widened the hole in the screen to make sure Billy would not scratch his fur when shimmying in!

As an incredible footnote, despite this experience, I still have a pronounced fondness for chocolate chip cookies.

—DR. FRANKLIN RUEHL, PH.D.

*D*r. Franklin Ruehl went on to share another personal experience with a telepathic cat named Simba. "I'll never forget my first encounter with him," Dr. Ruehl said. "It was when I was a graduate physics student at UCLA. Several of us shared a large office on the second floor of Kinsey Hall. I had just entered the office one autumn morning and placed my briefcase on my desk when I spotted a cat slinking toward me. He wore a coat of thick orange fur to which was attached a long tail ringed with alternating bands of black and orange. I was told that his name was Simba, and he greeted me with a lusty meow.

"I responded to his salutation by stroking his head, tickling him under his chin, and rubbing the sides of his soft body," Dr. Ruehl continued. "Simba acknowledged pleasure at such treatment by purring loudly."

Dr. Ruehl learned that Simba's owner, Ken, another graduate student, had been booted out of his apartment for smuggling a cat into his room. The graduate physics office was serving as a temporary refuge for the fugitive feline.

Having some knowledge of cats, Ruehl proceeded to gather some grass and leaves in a cardboard box and put Simba's forepaws in it, moving them back and forth in a scratching motion. The physicist knew that this was the technique that a mother cat uses to teach her kitten how to use a litter box, and it worked like magic for Simba until a formal sandbox was secured for him.

That afternoon, as Ruehl was perusing a physics journal at his desk, he noticed Simba scamper across the room and leap onto the table where the telephone was. The telephone rang a moment later.

"At the time, I did not pay any particular heed to the incident," Ruehl admitted. "But later that same day, Simba again went through a similar acrobatic routine, arriving beside the telephone just before it rang. Another student was in the room at the time, but he had not noticed Simba's mad dash across the room. The thought that Simba might be clairaudient flashed through my brain for the first time."

Two days later, Ruehl was in the office with Ken and an undergraduate who was seeking some assistance with a problem in quantum mechanics. "Suddenly," Ruehl recalled, "Simba bolted over to the phone milliseconds before it began to ring. This time, all three of us witnessed the occurrence. I asked Ken about it, but he swore that he had never seen Simba react in that manner."

During the ensuing week, Ruehl and several others observed Simba's telepathic behavior. Since the graduate students had been alerted to the potential phenomenon involved in Simba's dashes to the telephone, six additional incidents were recorded in a careful, scientific manner.

"I wanted to study Simba's apparent paranormal behavior under controlled conditions," Ruehl said. "But, alas, Ken had found a new apartment where pets were welcome, and he took Simba away from our domain. He told me that he kept a watchful eye on Simba, but he said that the feline never exhibited any precognitive aptitude in the new domicile.

"Could Simba's antics have been purely coincidental, with no psychic element whatsoever involved?" Dr. Ruehl asked, putting the query forward for examination. "Or did his behavior represent tangible proof of powers from another dominion?

"To accept the latter hypothesis, we must explain why

Simba manifested his telepathic powers only in the second floor office at Kinsey Hall. One possibility is that he might have been peculiarly sensitive to the vibrational frequency of incoming calls to the office phone, but to no others.

"Alternately," Dr. Ruehl hypothesized, "the specific construction of the office walls and the floor might have somehow amplified those vibrations and brought out Simba's latent psychic powers. While the walls and floor appeared outwardly to be composed of standard materials, perhaps some contaminants were present in sufficient quantity to render him telepathic while there.

"Or, conceivably, some type of psychic link existed between Simba and myself, whereby my presence gave his latent psychic abilities a 'jumpstart,'" the physicist speculated. "As far as I was able to ascertain, Simba exhibited his telephonic prescience only when I was also present in the room with him. While I personally favor this explanation, I can offer no tangible proof of its validity.

"While the enigma of his telepathic ability may never be solved," Dr. Ruehl conceded, "I will always remember my close encounters with a cat named Simba."

*O*n March 3, 1994, forty-
one days after the
Northridge, California,
earthquake, Tiffany, a ten-year-old Persian mix, was
found alive in the closet in which she had sought sanc-
tuary when the ground began to move under her feet on
January 17.

Laurie Booth, Tiffany's owner, had searched everywhere for her beloved cat. She had posted signs, taken
out newspaper ads, asked her neighbors in Saugus if
they could remember having seen the cat before or after
the quake.

Ironically, Tiffany had been only a few feet away from her desperate owner. Apparently, she had fled for safety to a neighbor's storage closet and had been accidentally locked in. When she was finally found, Tiffany was dehydrated, nearly starved, and only fleetingly conscious. She had had no food and only a little rainwater for forty-one days.

Veterinarian Sandy Sanford, who treated Tiffany at the Animal Clinic of Santa Clarita, put her on intravenous vitamin, sugar, and electrolyte solutions. Tiffany was fed orally every hour or so through a syringe because she was far too weak to feed herself.

"I still can't believe she's still alive," Laurie Booth told Rebecca Bryant of the *Los Angeles Times*. "When I picked her up she was just like a corpse. She was like a piece of tissue paper—frail and stiff and very cold."

Veterinarian John Burkhartsmeyer, owner of the Santa Clarita clinic, expressed his opinion that Tiffany had "just barely made it. . . . Bones and skin is all that's left."

Tiffany's inspiring story of survival and indomitable will touched hearts as far away as Europe. Veterinarians decreed her survival as "highly unusual, if not downright miraculous." Tiffany had become a symbol of hope to all those citizens of California who sought to rebuild their lives after the devastating effects of the quake of January 17.

On Friday, March 5, just as CBS News was at the

Santa Clarita clinic preparing to telecast a live report on Tiffany's fight for survival, the valiant cat went into cardiac and respiratory arrest. Veterinarian Sanford spent twenty minutes attempting to revive her with cardiopulmonary resuscitation, oxygen, and drug therapy, but Tiffany, who had lost about 60 percent of her body weight during her forty-one-day ordeal, was unable to respond.

Laurie Booth was distraught over her pet's death. How could she not have heard Tiffany crying for her when she was so close in a neighbor's storage closet? Why could she not have found her sooner so that her life would have been saved?

But Laurie remained thankful that she had been able to find Tiffany when she was still alive: "She didn't die alone in a cold place. She knew she was loved. . . . She definitely was a little fighter."

When Reverend Bob Short lived with his family in the high desert country of Southern California, he told his wife and children that the only time that they could keep a cat in their house was if they found one wandering homeless and helpless in that rough country.

"I had primarily directed such stringent requirements to our daughter," Bob told us, "because she would have turned our house into a menagerie if I had allowed it."

But it hadn't been very long after he had uttered the decree when members of the Short family happened upon a small baby kitten one night while they were out walking.

"It appeared to have been dropped accidentally by its mother on the way to its home," Bob said. "I took one look at this small creature and knew that we were going to be old and close friends. And as the kitten grew in size, it also grew very close to the family and more particularly to me. He would wait for me to arrive home, then he would jump up onto my shoulder and purr contentedly, as though he were saying, 'I am happy to see you home.'"

It wasn't long before the Short family began to notice that the cat could apparently read their minds.

"We would only have to look at him and ask if he were hungry or wanted to go outside, and he would either go to his dish or to the door. We also noted that when we wished him to come home for the night or for his evening meal, we had only to 'think' strongly on these matters, and within five to ten minutes he would come over the fence and arrive at our door—then cry out to be let in."

Kitty-Kat, as he came to be named, would sometimes ask to be let out shortly before the human members of the family wanted to go to bed.

"When it happened that he was still out and we fell asleep, one or another of our family would be awakened during the night by a dream of Kitty-Kat wanting to be let in at the door," Bob said. "It never failed when the dream so motivated one of us to get out of bed and

Cat Miracles ❧ ❧ ❧

investigate, Kitty-Kat would be standing there, waiting to be let in. It appeared that he could telepathically 'invade' our dreams with his desires."

One morning when Bob was sleeping soundly, he was greatly annoyed to be awakened by the sound of their front doorbell ringing loudly. And again. And again.

"I sleepily glanced at my alarm clock, saw that it was three o'clock in the morning, and staggered to the kitchen wondering who would *dare* to ring our doorbell at such an ungodly hour of the night.

"I flipped on the solarium light which illuminates the front door, then very warily stepped to the door and peered out. At first I saw no one, then I spotted Kitty-Kat standing there just before he started to cry out, as though to say, 'Well, here I am. Aren't you going to let me in?'"

Bob looked down at Kitty-Kat and said to no one in particular, "But the doorbell? What in the world?"

Bob squinted into the night. There was absolutely no one or nothing else that could have rung the doorbell. In the high desert country where they lived, they had no close neighbors. The darkness outside the small circle of illumination cast by the solarium light was silent. There were no sounds of jokesters running or driving away from their place.

Then he watched Kitty-Kat as he casually moved through the door and into the solarium. "Kitty-Kat," he asked incredulously, "just *how* did you manage to ring that doorbell?"

To this day, Rev. Bob Short told us that he doesn't know if Kitty-Kat somehow managed to jump up and hit the doorbell button several times in a row—or if the psychic cat managed to use mind over matter to ring the bell.

$\mathcal{W}$e are pleased and privileged to include in this book an inspirational piece by Janice Gray Kolb, author of *Journal of Love — Spiritual Communication with Animals Through Journal Writing*, in which she shares with us the special techniques by which she established spiritual rapport with her cat, Rochester:

God speaks to us within our thinking and in all of creation that surrounds us. To speak with nature, the birds, trees, plants and the animals is not a gift for a select few but is for everyone. We all have this potential if we only are patient and listen. But above all, we must love. And so if we are to communicate with an animal, we must love the animal and the divinity within this creature. Then the communication can begin if we give ourselves to it in love. With the same breath God created us, He created His creatures. Breath, life, and soul were known as *Anima* in Latin, and our word "animal" comes from this. We are told in legends about wise and holy people who communicated with animals and have lived peacefully and gently with Animalia, kingdom of the animals.

I realized soon after I adopted Rochester he was special. There are not adequate words to tell you how much I loved this precious marmalade-and-white kitten, named after the town in New Hampshire in which we met by God's appointment. But there was "something else" that made me know we were to have a beautiful life together. Three days after he came to live with us, Rochester and I were all alone one night on the sofa, and he came to me and bonded to me in a very unusual way that I have described in detail in two of my own books, *Compassion for All Creatures* and *Journal of Love*. Three times in succession he performed a little ritual of love—all done with his eyes holding mine, head bumping, and then circling around my shoulders, playing with my earrings only to begin anew. No sound was made, simply the meeting of our eyes, heads, and minds—and oh, I would add, hearts.

It was not an ordinary happening. This was something mystical. Something deep within myself assured me that Rochester and I were bonded in a most unique way. I would go over and over it in my mind, that such a tiny one repeated such a loving act three times and had communicated to me through this act and the love in his eyes that he had "somehow" made me "his."

Without my knowing that such a bonding is given by a cat when he has chosen a human, my tiny Rochester had made known to me, through his loving ritual and eyes that held me fast, that something on another plane was indeed taking place. He had even chosen a time of silence, when we were totally alone, for this marriage of hearts between kitten and human.

I felt that through this bonding he was trying to speak to me, to get my attention, and that he hoped I would be aware and very mindful. After this experience, I began to be even more sensitive to our relationship to try to be observant of all details. He was silent always and has remained so through the years. It is part of his mystical charm and contemplativeness. The only time there has ever been a sound is when he had made a slight "chirp" upon seeing me to show emotion.

His first choice in the early years was to speak to us with his body language, which he continues to do. He directs us with his eyes and actions and communicates to us in these ways. I always respond appropriately, never ignoring him, and my husband tries to respond also. Because our six children are grown and married, and we three live alone together in a small cottage on a lake in the woods of New Hampshire, we three pay attention to each other.

One amusing way Rochester used to ask if he could please go out on our screened-in porch, and if it is early in the morning and he is awake before we are, is to walk across the piano keys—full length. This was rather inventive! The piano in our cottage is right next to the door. He begins on the lowest keys and walks toward the door. It makes me smile because as a young girl I used to play a piece of music on the piano called "Kitten on the Keys." When we return home from anywhere, we hear him walking along the keys so he can be immediately inside the door to welcome us home.

For a very long time I felt inwardly that, though in each present moment we were sharing something extremely unique, more than the usual between woman and creature, I felt, too, that in time there would be more. I sensed there would be deeper communication, and I prayed for it and had an inner "knowing" it would happen. Each day there were new and subtle changes, and we grew in love and knowledge of each other. Now, many years later, he is still like a kitten in his running and playing and perfect health, and so much love and devotion has transpired between us. He is my feline soul mate.

And our communication truly has deepened immensely.

We can communicate without spoken words. Mind to mind. I know he knows what I am saying, and I know what he is telling me in his love, expression, and silence. And when I act on certain things that I believe he is communicating, they prove to be so. But I also speak to him aloud, treating him as my precious friend, and sometimes carry on long conversations, which he will attentively listen to. Or I will ask him his opinion and believe I receive it if I have been pondering a matter. I talk to him and treat him as if he can understand, giving him this dignity and respect. Many beautiful and intimate moments have transpired between us because of this.

The best way to capture moments is to pay attention in an open way, to attempt to practice mindfulness in life. I have become so much more aware now in recent years to observe Rochester's actions in response to my words said aloud to him in normal conversation. His responses are not out of habit. I am constantly changing what I suggest to him to test his understanding of my words, and then I let him lead if it involves action. He leads me through his unspoken mental words and his expressive golden eyes.

If I had talked to him and merely said the words and paid no attention to his responses, I would have missed these daily "replies" through his actions and not seen the wonder of it all. Awareness and appreciation for the present moment are necessary, and through these will come mysterious blessings.

But mind to mind we also talk, and so bodily action and response are not always necessary on his part. We can be sitting together, he on my lap, and I talk to him mentally, believing I know his replies. The conversation can continue at length. It can also occur when he is near by me as I write or when I am out in the car and he is at home. I send him mental love notes and images. I believe with all my heart I receive them back. There are so many things in daily life that tell me we truly communicate mind to mind, heart to heart. I can never doubt that no matter what others may say or how foolish they might try to make me feel, for anyone who knows me personally knows my undying love for Rochester. Others might feel I am making all of this up or am delusional. It is not so. And after writing my own book on this subject of communication, I know there are others out there in the world who also experience this with their beloved companions and have written books also. Since *Journal of Love* was published, I have received so many beautiful letters and e-mails from readers who have shared their own wonderful experiences in animal communication. I received a long e-mail from a woman in the United Kingdom telling me how much the book and Rochester meant to her. A blind woman e-mailed me to say that *Journal of Love* was teaching her to communicate with her seeing-eye dog.

Often I thought to write down Rochester's day-to-day experiences in a journal I would keep separately for him alone. I have kept journals of my own since I was a little girl. That is how much credence I gave to believing I knew what Rochester thought and was saying to me! Since he was always there for me, constant and ever present, no matter what went wrong in my own life, I did not think it unusual then to consider keeping a separate journal for him as well as continuing to record his precious life in my own journals. It was inevitable because of my nature and also because we seem so much alike in personality. I knew from our mental conversations this would be pleasing to him. He has spent his entire life watching me write for hours each day, and I felt now it was his turn. Since he does not speak aloud, it seemed to me that writing his thoughts for him was the thing to do. I felt that they were worthy thoughts and that I should have written them down for all time, to read and reread. A journal keeper thinks in terms like this.

His presence and his mental thoughts in response to mine had nourished me spiritually and emotionally for years. How much more to have his thoughts in writing to bless me? Since I really love Rochester, this was the natural progress of events in our communication.

What I had heard mentally before as we carried on conversations in silence—or when I would speak aloud and I would receive a mental answer or one of action through body language—became different when I began to write. When I finally obeyed and put pen to paper, there was no doubt it was Rochester speaking to me. There was a different sort of energy. We seemed held in an aura together. I would ask him to speak, and he would watch me poised to write, then he would begin, and I would write as fast as I could. At times I asked him a question first. Other times he read my mind.

Even though I have experienced this now so many times, I am still and always filled with astonishment and gratefulness also, to all that is divine that is creating this precious and awesome communication. My hand and pen move, taking dictation that comes through me from him.

I have shared my own spiritual experiences of communicating with Rochester, but I believe that everyone has this gift of being able to communicate with their companion animal and then to go on and communicate with other animals, too, as I have been able to do on occasion. There is much to learn.

Many reading may already communicate with an animal friend. In my own book on this subject I have written out a step-by-step formula for communicating with your animal companions through writing. I will state it in a briefer form here:

1. Find a quiet place.
2. Have a pen or pencil and paper with you.

3. Pray and ask God and His Angels to help you.
4. Meditate. (This will bring a calmness to you and your animal. Meditate briefly.)
5. Talk to you animal companion.
6. Pick up your pen and write. (Write any word or sentence that comes into your mind. The more you write, the more will come. Do not stop to read any of it until you feel as if the message is complete. You will have a "knowing." Then you may reread.)
7. Do not doubt.
8. Say "Thank you." (Thank God and His Angels for their help and presence. Thank your beloved animal companion for the words he gave you.)

The more you practice with your friend, the more confident you will become. If you prayed, then expect the messages to be from your companion. Do not take credit for words that are not yours.

And then one last significant thing—

9. Buy a special journal or notebook. (If you use an attractive blank book you are giving credence and importance and sacredness to what you are about to do and record. This indicates you believe and are expectant and that you treasure this communication that is about to happen. A special journal or notebook confirms and also honors your animal

companion. You then will forever have all the messages safely kept to reread—a piece of paper would eventually be lost. This builds your faith. Be sure to date each message or conversation.)

Once you see words written and on paper in a journal for all time, I believe that writing your companion's messages will be your most desirable and cherished form of communication. Rochester and I have grown so deeply bonded in love through our communications. How I wish I had known this before with other animals that were a part of my earlier life.

I feel certain this interspecies communication is how God wished things to be in His world. May God and the Angels bless you and your companions.

—JANICE GRAY KOLB

Authors' Note: It is with great sadness that we must report that Rochester's beautiful spirit entered the Great Mystery on Friday, March 8, 2002 at 5:07 P.M. Jan tells us that, "I am richer for having been taught and loved by Rochester. Since his passing, I have had ongoing experiences with his presence, and he has given me gifts of spirit communication that I will have until we are together in Heaven."

*D*r. Franklin R. Ruehl, a nuclear physicist and author who is well known for his articles on science for the layman, has often pondered whether or not humans can actually establish a psychic link with animals. "Is the concept of interspecies telepathic communication totally illogical?" he has asked as a lead-in to his popular Los Angeles–based cable television program, *Mysteries from Beyond the Other Dominion.*

"An affirmative answer appears applicable to the former interrogative," he concludes, "based on the results of an extensive battery of experiments conducted by Duke University researchers back in 1978."

Dr. Ruehl went on to explain that Stuart Harary was an undergraduate student majoring in psychology at Duke who claimed to be able to project himself into an out-of-body experience (OBE) at will. In order to test him, he was placed in either of two rooms in Durham, North Carolina — one at Duke and the other in Building A at the Psychical Research Foundation (PRF). A kitten that had previously exhibited a rapport with Harary was placed in a target room in Building B, which Harary's out-of-body spirit was to visit. Building A was some fifty feet from Building B and a quarter of a mile from Duke University.

Researchers had noted that normally the kitten, which was just seven weeks old, was calm and quiet when Harary was physically present in the same room. Whenever Harary left the room, the kitten would become noisy and restless.

For the experiment, the kitten was placed alone in a wooden box with its floor divided into twenty-four numbered squares. The behavioral measures employed to assess activity were the number of meows per 100 seconds and the number of squares entered into per 100 seconds by the kitten. The kitten was "scored" as entering any square into which it had placed both forepaws.

"Intriguingly," Dr. Ruehl said, "when Harary was

placed in one of his experimental rooms and was in the process of undergoing an out-of-body experience, the kitten did not meow once in eight different OBE periods. It was as though Harary's calming spirit had entered the room with the cat. However, during eight control periods when Harary was in an experimental room and *not* undergoing an OBE, the kitten meowed thirty-seven times. Its 'meow rate' was 0 per 100 seconds during OBEs; 3.85 per 100 seconds during controls."

Dr. Ruehl also pointed out that the kitten became extremely inactive during an OBE, averaging only 0.21 squares entered per 100 seconds, as opposed to 3.54 per 100 seconds during a control. "Once again," the physicist noted, "the kitten appeared to have sensed Harary's presence during OBEs."

In another phase of the experiment, Harary was asked to *pretend* he was undergoing an OBE by trying to imagine himself in the room with the kitten, patting it and playing with it. In other words, he would go through the mental motions of having an OBE without actually experiencing one. "Significantly, the kitten was not calmed during the imaginary OBE," Dr. Ruehl explained.

In yet another stage of the investigation, the kitten's orientation relative to Harary's spirit's position in the room was measured. Harary visited four different locations, fifteen feet apart, each near the kitten. "Importantly," Dr. Ruehl emphasized, "the animal displayed a definite shift in position toward Harary's actual OBE locations."

In Dr. Ruehl's opinion, "This dramatic demonstration of a psychic bond between man and cat is all the more impressive when one considers that the kitten was an absolutely objective participant—with no bias whatsoever as to the outcome of the investigation. However, further testing of the kitten became impractical when it became accustomed to the experimental box and the researchers handling it.

"As a fascinating sidebar," Dr. Ruehl added, "it should be noted that many of the human participants also perceived Harary's spirit form when he was projecting his spiritual essence during an OBE. This experiment must be ranked as one of the most impressive experiments in the annals of parapsychology."

*A*ll true cat lovers will be delighted to regale any listeners within earshot with an endless supply of anecdotes that demonstrate conclusively the intelligence, resourcefulness, and all-around wonderfulness of their feline companions. The love affair so many of us have with the cat goes back to ancient Egypt, at least 2,500 years ago. At first the relationship between humans and the small wild cats they managed to capture from the surrounding forests and deserts was one of mutual benefit. The citizens of Egypt found a dependable animal that would keep their precious supplies of grain free of rodents, and the cats that once had been surviving on their own found a reliable source of shelter and food.

From ridding the granaries of mice and rats, the cats soon graduated to guarding the Egyptian temples, and it wasn't long before the once-utilitarian rodent catcher had been transformed to a deity. The Cat-Mother goddess Bast (or Ubasti) became associated with the benevolent aspect of Hathor, the lioness, who was said to have nine lives. Interestingly, the Egyptian word for cat was *Mau*, which can be heard at once to be an imitation of the animal's call and the nearly universal human infant's cry for mother. The Cat-Mother became worshipped with such intensity that the wanton killing of a cat was punishable by death.

Because the Egyptians had a great fear of the dark, they observed that the cat, a nocturnal creature, walked the shadowed streets after nightfall with great confidence. Carefully evaluating the meaning of the cat's nighttime meanderings in association with the benevolent Bast, the Egyptian sages deduced that the cat was solely responsible for preventing the world from falling into eternal darkness.

From ancient Egypt to the twenty-first century, the growing bond between felines and humans has now extended to include almost the entire planet. In the United States alone, there are approximately 63 million pet cats—all gracing some proud owner's windowsill, couch, or kitty basket. And as crazy as we Americans are for cats, our friends "down under" own more cats per capita than we do. According to a chart published in *Petfood Industry* magazine, Australians rank as the Number One cat-lovers in the world, with 33 percent of their families owning at least one cat. The United States follows with 30 percent of all households sharing their space with a cat. Filling out the top slots are Canada and Belgium, tying for third position with 29 percent of their populations cherishing a relationship with a cat; and France and Switzerland sharing fourth position with 27 percent of their homes boasting a feline member. Japan is at the bottom of the chart with only 5 percent of their families keeping a cat for company.

Those who cherish their unique relationship with their cats are quite aware of their merits, but it seems appropriate at this point in the book to share a few interesting bits of research to demonstrate how remarkable those feline friends truly are.

Like most cat owners, you have probably wondered exactly how your pet is able to make that soothing purring sound, but you've probably never wondered much about why it is that your cat purrs when it rubs against your leg or when it closes its eyes contentedly as you scratch behind its ears. You've probably just assumed that those were the unique kitty sounds of love and affection.

The sound of purring is unique to cats, and with the exception of tigers, all sizes and shapes of the feline family purr. But the melodic sound of a cat purring represents far more than feelings of contentment, well-being, or affection. In March 2001, scientists from the Fauna Communications Research Institute in North Carolina released their findings that the purring of cats is actually a natural healing mechanism that throughout the centuries helped to inspire the myth that cats have nine lives. When a cat is wounded or injured, it purrs because the sound frequencies thereby emitted help their bones and organs to heal and to grow stronger.

Dr. Elizabeth von Muggenthaler, the president of the research institute, compared the healing process involved in a cat's purring to the effect of an ultrasound treatment on humans. Exposure to certain sound frequencies is known to improve bone density in humans. Similarly, the purring mechanism in cats somehow creates sound waves at a particular frequency that triggers the healing process in feline bones and organs.

Dr. von Muggenthaler also commented that purring had to be advantageous to cats in order for it to survive natural selection and that there seemed no obvious advantage for a cat to purr merely to exhibit contentment. Next on the agenda for the research institute,

Dr. von Muggenthaler added, would be to attempt to explain the mechanics of the purring process.

And then there are all those discussions about how far a cat can fall and survive. It seems Mogadon, the supercat of Leeds, Alabama, described earlier in this book, who fell twenty-one stories to the sidewalk and lived, is only one of many.

A recent study published in *The Journal of the American Veterinary Medical Association* found that out of 132 cats that fell an average of 5.5 stories, 90 percent survived, including one that fell 45 stories.

Research has determined that a cat can reach a terminal velocity of about 60 mph after free-falling about 130 feet in a few seconds. In actuality, the greater the distance a cat falls improves its chances for survival. In one study, of the twenty-two cats that scientists documented who had fallen eight or more stories, only one had died. According to an extensive study, a cat falling a great distance will have more time to relax and to position itself for minimum injury. Cats appear to accomplish this by arching their backs, twisting their torsos independently of their hind legs, and then bringing their hindquarters around. By thus spreading their limbs in a horizontal position, much like the position assumed by the so-called flying squirrels, cats are able to distribute the points of impact fairly evenly throughout their entire body. By contrast, falling humans will reach a terminal velocity of 120 mph after plummeting for a few seconds, thus greatly limiting the height from which they can survive a collision with earth or concrete.

Research has indicated that extremely high notes may cause agitation in many cats because the sounds may replicate certain "words" in the special language of felines and may even approximate tones similar to those expressed during the courtship ritual. In other cases, very high notes may imitate the cries of a distressed kitten and may especially upset adult female cats.

The very impulse to pursue rodents may have its basis in a cat's musical sensitivity. Biologists have discovered that it is quite likely that all mice sing in two-octave ranges and tempos, varying between two and six notes per second, producing songs similar to the chirping and twittering of small birds, but with a great deal more variety. Humans are seldom aware of such rodent chorales because the vast majority of mice sing far too high for the human ear, perhaps much like the supersonic squeaks of bats. The mice that people do hear may be compared to the basses and baritones of a human choral group, while most of their fellow warblers are very high sopranos.

On the other hand, cats, with their remarkable supersensitive hearing, present a captive audience to the pesty little chirpers, and some scientists have theorized that cats receive an additional incentive to pursue mice when the little buggers sing off-key and upset the feline insistence on harmony. Imagine the terror of those unfortunate miniature singers when they receive the ultimate negative review of their performance from a cat with a penchant for perfect pitch.

When it comes to size, as far as we can determine, the crown for the world's heaviest cat graced the head of Himmy, an Australian cat that tipped the scales at forty-five pounds, ten ounces in 1982.

In 1991, the tabloid newspaper *National Enquirer* conducted a contest to determine the heaviest cat among those owned by its millions of readers. The winner was Spike, a thirty-seven pound, thirteen-and-a-half-ounce tabby owned by Gary Kirkpatrick of Madrid, Iowa.

In 1992, the *National Examiner's* Fattest Cat in America contest located Morris, a tubby tomcat owned by Fred and Jeannie Scott of Ottawa, Kansas. Morris weighed in at a hefty thirty-six pounds. The runner-up in the competition was Tiger, a thirty-three-pound brown tabby belonging to Paul and Teri Hammer of Excelsior, Minnesota.

The world's smallest breed is the Singapura, a street cat that lives in drains in Singapore. Although quite rare, this small, short-haired feline with large eyes, an unusual beige and brown ticked coat, averaging only about four pounds in weight, has found its way to the United States in recent years.

Folklore and ancient myths alike attribute nine lives to a cat, but just how long do cats manage to stretch out a lifetime? Since the 1930s, the average life expectancy of cats has nearly doubled, from eight years to sixteen.

The longevity champ appears to be Puss, who passed away in 1939 just one day after the celebration of his thirty-sixth natal anniversary.

The runner-up is quite likely Grandpa, a hairless cat that belonged to Jake Perry of Austin, Texas. Grandpa dined on bacon and eggs, broccoli, and coffee with cream every morning for breakfast until he died at the age of thirty-four in 2001.

Spike, a ginger-and-white tomcat, was pressing the record when he died on July 11, 2001, just two months after his thirty-first birthday.

As of January 2002, according to *The Guinness Book of World Records*, the oldest living cat was Bluebell, twenty-four, owned by Lea Yergler of St. Maries, Idaho.

On the morning of July 17, 1992, Estelle Littmann of Montgomery, Alabama, was in her station wagon on her way to the bank where she had worked as a teller for more than eleven years. She was only a few miles from her home when she was startled by a man in a brown van who pulled up beside her and began to shout and wave wildly.

Later, Estelle admitted that she had been frightened. The media often carry accounts about killers and maniacs trying to assault women on the highways. So at first she tried to ignore the shouting, waving man in the van.

But the man pulled ahead of her, flashed his signal lights, and motioned for her to pull over.

Estelle thought no way was she going to pull over and become the next victim of some serial killer. She floored the accelerator and shot around him at top speed. She did, in fact, drive about fifteen miles over the speed limit, hoping that some patrol car would spot her and come to arrest her—and save her.

When her pursuer caught up to her once again, Estelle Littmann had another thought. Perhaps this was all connected to her position with the bank. She remembered seeing a movie about some bank robbers who first kidnapped bank personnel and then forced them to assist in the theft.

At last she spotted a housing development where a security guard was stationed. She felt certain that if she drove up to the guard on duty, the nutcase following her would flee.

As she pulled into the driveway by the guard's station, the man in the brown van honked his horn several times and drove on.

Estelle thought the persistent pervert had given her one last beep on his horn, but she didn't care if he thought he was insulting her or not. At last she was rid of him.

The security guard stepped out of his station to ask what she wanted, and she watched his face turn pale.

"Ma'am," he blurted out, "you got a cat on the top of your car!"

Estelle quickly unbuckled her seat belt and pushed open the door of the station wagon. Her mouth dropped open, and she knew she was turning several shades paler than the security guard.

There, on top of her station wagon, spread-eagled on the roof and clutching onto the luggage rack for dear life, was her black-and-white tomcat, Ronald! The poor cat had just had the ride from Hell, and he looked as though he was frozen solid in complete and total fear.

Estelle Littmann said it took two or three days of lots of tender loving care to "unthaw" Ronald, but he was soon doing fine, none the worse for his ride of terror on the top of a speeding car.

*M*arty probably got himself into trouble while engaged in his favorite pastime of chasing mice into holes around the patio of the Dunbars' apartment building in Edina, Minnesota.

What most likely occurred on the afternoon of April 9, 1992, is that Marty felt a little shy when the maintenance men came around to work on the patio, and found a hole big enough to hide his yellow-striped body. Of course, Marty had no idea that the workmen would place a ten-inch slab of patio concrete over his hiding place.

When Marty had not come home after three days,

eight-year-old Frankie Dunbar was in tears. He had received Marty on his fifth birthday, and they had become fast buddies. Over and over he sobbed his anxiety to mother, Robin, asking the question that she could not answer: "When is Marty going to come home?"

On the fourth day after Marty's disappearance, Greg Dunbar decided that the cat's disappearance was probably something more serious than the unannounced holidays that cats are prone to declare. As soon as Greg returned home from work, the three members of the Dunbar family scoured the neighborhood, calling for Marty. Later, after dinner, they posted reward notices in all the supermarkets within a reasonable cat-hike from their apartment building.

On the evening of the ninth day, Robin and Greg sat Frankie between them on the sofa and explained that just maybe Marty had had an accident and wouldn't be coming home. Frankie cried himself to sleep.

Eleven days later, on Easter Sunday, Marty appeared at the Dunbars' patio door covered with dirt and his paws so muddy that it looked like he was wearing miniature boxing gloves.

"We were astonished," Greg admitted. "Speechless. We could only blurt out half-sentences like, 'Look there . . . there he is . . . Marty. Look. Marty has come back.'"

Frankie was overjoyed to see his pet, but Marty didn't stand on ceremony or have too much time for tearful reunions. He headed straight for his food bowl and began to eat, apparently trying to gain back in one meal the weight he had lost in eleven days.

When Greg and Jerry Faciana, the maintenance supervisor at the Dunbars' apartment complex, found the hole in the lawn, they began to reconstruct the circumstances of Marty's disappearance and his return. It seemed likely that Marty must have crawled into a hole near the patio on the day when the workmen were patching the area with new stones and concrete. Marty probably hid from them, and they poured a slab of concrete right over the place where he lay crouched in a hole.

"Once he realized he was trapped," Greg Dunbar theorized, "Marty probably started trying to dig himself out. We figure he must have stayed alive by burrowing into pockets of mice as he dug."

Jerry Faciana commented about the condition of Marty's paws. "The poor guy didn't have any claws left at all. He had worn them down to nothing from scratching his way to freedom."

Later, Frankie asked his parents whether Marty had used up all of his nine lives in his escape. His father answered that Marty had always been a fighter.

"I'll just bet he saved at least one or two lives for future emergencies," Greg Dunbar said.

It certainly appeared that Mouse had used up all nine of her lives when the Wilkerson family of Invergordon, Scotland, buried her in August 2001. Their daughter Sarah had come running into her parents' home carrying the body of a black cat that had been struck and killed by an automobile. Mrs. Lou Wilkerson recalled that Sarah was inconsolable when she came into the house carrying the battered body of Mouse.

Sarah wrapped Mouse's body in a tea towel and laid it on the sofa. She caressed the cat tearfully and tenderly, then went home to try to regain her composure.

Mrs. Wilkerson placed Mouse in a shopping bag from Harrods department store and decided that the dear cat should have a dignified funeral service. The Wilkersons kept a small garden, and she knew that it would not do to bury Mouse in a plot where she would continually be disinterred, so she prevailed upon a friend who lived thirty miles away in Edderton and who had a bit more land and could provide a decent burial site for the cat. After all, Mouse was a cat of great distinction and was much loved by many people, so she deserved a proper place of eternal rest.

After a touching service of remembrance, Sarah returned to her own home to be alone with fond memories of the beloved Mouse. She had not been too long into her reverie when Mouse came bouncing into the kitchen, mewing for her afternoon tea.

Sarah blinked her eyes in astonishment. The creature that she saw before her was neither a ghost cat nor a zombie cat. It was the flesh-and-blood Mouse.

Sarah was, of course, ecstatic that her beloved Mouse was not dead after all, but she was rather embarrassed to notify friends and family that there had been a case of mistaken identity. The cat for whom they had held such a lovely burial service was not Mouse, but an unidentified feline whose spirit had been blessed with a loving and tearful farewell.

One of the most common sayings about cats is that they seem to possess nine lives because of their ability to emerge relatively unscathed from the most desperate or deadly of circumstances. A number of the stories that people have shared with us about their feline pets would nearly have us believing that a cat really does have more than one life to draw upon when it gets into situations from which it would seem impossible for any living creature to survive.

In January 2001, Minnesota resident Roberta Johnson was making a left-hand turn in her automobile off a cold winter's highway when she glanced out her side window and saw a large block of ice with a feline face inside. Because she had rolled down her window to signal the turn, she saw the poor little face clearly and assumed that the creature was dead, frozen solid inside the chunk of ice. Then she heard a "meow" issuing from the pathetic kitten and was astonished to consider that the cat might still be alive.

Roberta pulled over to the side of the road and went to examine the block of ice. Upon closer inspection, she realized had obviously fallen from a car's wheel arch. How the cat had survived such a traveling experience almost subtracted from the impact of the greater question of how it had survived frozen in the ice that had formed under the arch. But survive it had. Outside of frostbitten ears, the cat was fine. Roberta Johnson, not one to turn her back on the strange destiny that brought them together, decided to keep the cat and name him, Car Cat.

In August 2001 in Herning, Denmark, a twelve-week-old cat named Sylvester jumped into a washing machine just before it was turned on. The fortunate feline was dizzy, blue, and yowling his head off, but he was alive after surviving a twenty-minute, seventy-degree C tumble before his owner Bianca Marten was able to rescue him.

Bianca took Sylvester to a veterinarian to be certain

that he was all right. After listening to her story and examining the cat, the vet proclaimed Sylvester's surviving the wild ride in the washing machine to be a miracle and pronounced him as good as new. Sadly, though, Sylvester must still have been more than a bit dizzy, for he crawled back into the washing machine as soon as he and his owner returned home. However, Bianca Marten said she would henceforth always check for a stowaway cat in the washer before turning it on.

On December 11, 2001, a cat named Gurbe survived being trapped in a tumble dryer set at seventy degrees C for ninety minutes. His owner, Metha de Bruin of Terband, Netherlands, didn't know what had happened until she began noticing cat hair in the filter of the dryer. Opening the door to the machine, she found Gurbe between the sheets, which were splattered with blood.

Metha ran to a neighbor, fearing the worst for Gurbe, assuming that he must be dead. By the time she returned to her home with her friend, however, Gurbe had managed to crawl out of the dryer. Later, a veterinarian declared him to be fine, just a little dazed and a bit bloodied from the experience.

An eight-week-old tortoiseshell kitten, named Flowerpot after the contents of the crate in which she was found by stevedores, managed to survive for more than a month inside a box shipped from Penang, Malaysia, to Salisbury, Wiltshire, UK. A representative from the Royal Society for the Prevention of Cruelty to Animals theorized that Flowerpot had stayed alive by licking condensation that formed on the inside of the crate.

Bessie survived seven weeks in a chimney in Shoreham, Kent, by drinking rainwater. Apparently the cat had fallen down the chimney while on a neighbor's roof, landed on a ledge, and was unable to climb back out. Long considered missing by her owner David Hutchins, Bessie eventually became so weak from lack of food that she fell off the ledge and landed in Sarah Philip's fireplace on October 6, 2000.

A ten-week-old cat arrived in Peterborough, England in late August 2001, after traveling 2,000 miles over land and sea for seventeen days without food or water. The container had been sealed in Tel Aviv, Israel, without knowledge that a cat was inside. Named "Fizzy" after being discovered during the unloading of the container at the Soda Stream factory, the cat had only condensation to drink during the long trip.

On November 2, 2000, a very fortunate cat named Anthony was recovering in his owner's home in Seehausen, Sachsen-Anhalt, Germany, after being trapped for twenty-six days between two walls without food or water.

Polly was little more than skin and bones but happily back in the arms of her owner, Maureen Morris, after surviving six weeks trapped under the concrete floor of a newly constructed house in Clacton, Essex. Ms. Morris had left notices everywhere, including the site where the new house was being built, but she had given up on ever seeing her beloved cat again.

Fortunately, in late September 2001, a couple viewing the property said that they had heard a cat mewing somewhere in the house. When the construction company sent someone to investigate, a hole was made in the floor to liberate Polly from the place where she had been imprisoned for six weeks.

On December 8, 1989,
Rhea Mayfield of
Brownwood, Texas,
asked her daughter to help her move the coffee table
from their apartment to the building's storage room to
make space for the Christmas tree. When they returned
to the apartment, they noticed that Kelly, their tabby,
was missing. After several minutes of calling for her, they
concluded that Kelly must have wandered out of the
apartment while they were lugging the table to the
storage room—but it never occurred to either one of
them that she could have curled up for a nap inside a

compartment in the coffee table.

Rhea called the police and placed ads in local newspapers. Every day she would allot some time to searching the neighborhood for her missing cat. Anticipation over the approaching holiday season was replaced by anxiety over Kelly's welfare.

It was not until January 22, 1990, forty-six days later, that the manager of the apartment building happened to hear the weak cries of a cat while he was in the storage room. It took him a few moments to locate the source of the pathetic cries, but he opened the compartment of the coffee table to discover a very emaciated cat, barely recognizable as Rhea Mayfield's robust and slightly plump Kelly.

Kelly barely had the strength to purr against her owner's cheek, so Rhea took her directly to a veterinarian. Here it was determined that the cat that had previously weighed in at seventeen pounds, now tipped the scales at less than five. A professor of veterinary medicine reviewing the case remarked that a dog would not last a week without water, and the longest that he had ever heard of a cat surviving without liquids was thirty days—never forty-six! He theorized that the temperatures in the storage room, sometimes dropping several degrees below freezing, might have helped to save Kelly's life by slowing down her body functions. Whatever the circumstances that enabled her to survive, Kelly is definitely a super cat.

*L*aura Dalfonso was heart-broken when Smokey, the smoky gray Siamese that she had had for a little over a year, disappeared in July 1990. Laura owns a vending machine business in a suburb of Baton Rouge, Louisiana, and she had gotten into the routine of taking Smokey everywhere with her on her rounds. He sat on a pillow next to her chair when they watched television, and he rode on her lap whenever they would go anywhere in the car.

Laura spent the days after Smokey disappeared searching the neighborhood and tacking up reward posters in all the malls and supermarkets.

Then she got to thinking that she had sold six vending machines to Martin Napier the day Smokey had disappeared, and she started getting a feeling that Smokey might have somehow gotten inside one of those machines.

Napier was sympathetic when she phoned him. Although she said she doubted Smokey could actually be hiding in a vending machine, Napier promised to go out to his warehouse and check them right after he hung up.

He did check them out as he had promised, but he didn't open them up and look inside. After all, he had reasoned, every slot and opening was closed, so he couldn't see any way on earth a cat could have wriggled between a hairline crack and gotten inside. In addition, he had placed his ear to every one of the machines to listen for anything that sounded like a cat, and there were no strange sounds coming from any of them.

Laura Dalfonso continued her lonely quest, taping up hundreds of LOST CAT signs and walking through the woods and nearby neighborhoods calling Smokey's name. After thirty days, she gave up hope that she would ever see Smokey again.

A few days later, Martin Napier was beginning to prepare his vending machines for service. A local school board had bought six, and it was almost time to install them for the fall term.

Napier said later that someone could have knocked

❧ ❧ ❧ *Cat Miracles*

him over with a feather when he opened up the fourth machine and saw those big blue eyes looking up at him from a scruffy cat face. It was Laura Dalfonso's cat, Smokey. Somehow he had managed to get himself hidden away in a compartment that was nineteen by thirty inches—and only eight inches high.

Napier put his hand out, and Smokey rubbed his head against the man's palm. It was obvious that the cat was too weak even to stand up. Napier rushed Smokey to a veterinarian, who told him that the cat had probably lost about half his weight through dehydration and malnutrition. The veterinarian admitted that thirty-seven days without food or water was an amazing feat of endurance, but she added that cats seem to have the ability to go for long periods without nourishment of any kind.

When he was certain Smokey would be all right, Napier called Laura Dalfonso and gave her the good news. She had been right about Smokey being in one of the machines, he told her, but he was going to be all right even after thirty-seven days of solitary confinement— without bread and water.

Smokey made a full recovery, and he returned to Laura, to go everywhere at her side once again.

On that very busy day
in 1990 when the
employees of the freight
company in Kent, England, were sealing the metal
container that was to protect the Mercedes-Benz
during its sea voyage to Australia, they had no idea
that they were also enclosing a stray black cat inside
the box. The unwilling feline stowaway—who came to
be called "Mercedes"—survived an astonishing fifty
days without food or water, locked securely in its metal
seagoing tomb. In an unparalleled feat of endurance,
Mercedes traveled 17,000 miles in her metallic crypt;

and when the ship arrived in Port Adelaide, Australia, nearly two months later, customs officials were stunned when the skin-and-bones cat stumbled out of the container.

Dr. John Holmden, a veterinarian and chief animal quarantine officer in south Australia, theorized that Mercedes must have had a full stomach before she became trapped in the container. By licking drops of condensation and by spending nearly all of her time resting, she managed to stay alive.

Under Australia's strict quarantine laws, Mercedes had to be detained for nine months. However, the quarantine was not all bad, for she was fed generous amounts of cat food and milk and soon bulked up from the four pounds she weighed when she tumbled out of the container to about eight pounds. And after her confinement, the owner of the Mercedes-Benz said that she intended to keep the feisty cat who had shared the metal container with her automobile from Kent to Port Adelaide.

Carol Gringas of Middleboro, Massachusetts, returned from a Florida vacation in the winter of 1990 to find her five-month-old kitten, Snowball, skewered through the head with a two-foot arrow. Although Carol and her husband had left their white Angora under someone's supervision, some fiend had fired upon Snowball when he was out of the house.

A veterinary surgeon performed three hours of delicate surgery to remove the two-foot-long, quarter-inch shaft that had pierced Snowball's nostrils, traveled through his sinus cavities and throat, and exited through the back of his neck. The arrow was one and a half times as long as the kitten's body and thicker than his leg bones. If it had struck a fraction of an inch above or below where it did, it would have pierced the brain or the spinal cord and undoubtedly have killed him. After a second operation and a month's stay in the animal clinic, Snowball made a miraculous recovery.

Apparently there are some irresponsible archers running amuck who take perverse delight in using cats for target practice. Zena had been missing from her home in Carlisle, Ohio, for about three weeks in the winter of 1993 when she suddenly returned with an eight-inch arrow protruding from her head. Although Zena didn't seem particularly perturbed by the peculiar circumstances, her owners rushed her to the nearest veterinary clinic.

Judging from the depth of the wound, the veterinarian determined that whoever the cruel archer had been, he or she had released the arrow at point-blank range. Since Zena had already used up at least one of her lives and didn't seem at all agitated about the arrow protruding from her forehead, the veterinary staff allowed her to rest a few days in order to permit the wound to heal and to avoid internal bleeding. Zena was purring and acting normally, except she had difficulty eating or drinking from a bowl with the arrow in her head.

After he had performed the delicate surgery, veterinarian Ray Ruhrmund commented that if the arrow had struck Zena just a half an inch higher, it would have entered her brain and killed her immediately.

On October 31, 2001, RSPCA Inspector John Bowe said that he had never seen so many pellets in an animal that had taken the blast of a shotgun at close range and lived. A veterinarian surgeon picked forty pellets out of the riddled body of Fizz, the pet of the Pamphilion family, after the three-year-old black-and-white cat managed to limp home to his owners in the village of Briston in north Norfolk, England after having been missing for four days.

Two-year-old Macey fell from a window in Edinburgh, Scotland, on June 26, 2001, and impaled herself on the spike of a rusty railing some thirty feet below. Firefighters cut through the spike and brought the cat to the veterinarian with the metal still protruding from her leg. After a seventy-five-minute operation, Macey was said to be doing very well.

When Bob Perciaccante of Fredericksburg, Virginia, first saw the half-frozen bedraggled cat on Christmas Day 1993, he figured that it had already used up several of its nine lives. His roommate, Jennifer Anderson, spotted the shivering kitten near the garbage cans of their apartment building, and she and Bob decided to befriend the feline and take it home with them. For perhaps obvious reasons, they named the ice-covered kitten, Slush.

Bob and Jennifer had scarcely nursed Slush to a higher degree of physical health when he disappeared for a few days—then returned to them with his jaw blown away. The unfortunate feline had apparently aroused the ire of someone with a gun. Since the cat could not eat or drink with his lower jaw hanging off, the two roommates pooled their funds and took Slush to a veterinarian clinic to get repaired.

After barely a week back home in the apartment, Slush ran across the street at a most inappropriate moment and got hit by a car. Somehow he survived the ordeal with only an injured leg, but while he was hopping around unsteadily on three legs, Slush fell into a pond and nearly drowned. A neighbor happened to see the cat struggling in the water and was able to pull Slush out before he drowned. Once again, there were veterinarian bills to get Slush back to health.

So how many lives does Slush have? Bob and Jennifer can account for quite a few, for they have witnessed him frozen and starved nearly to death, shot in the face, run over by a car, and almost drowned. But they disregard the financial sacrifices they have made to keep their feline friend alive, because they consider Slush to be the most affectionate cat in the world.

*P*erhaps there is one cat that possesses even more than nine lives. In early February 1991, during the peak of the chaos surrounding the Persian Gulf conflict, rumors abounded that Iraqi terrorists had pinpointed a number of United States cities for reprisals. Law enforcement agencies around the country were extra vigilant, and in a major southern city, a hapless kitten was mistaken for a terrorist bomb and detonated.

At around 6:00 A.M. on that February morning, Sheriff Charles Herbert received a report from two patrolling deputies that they had spotted a mysterious box in the driveway between the county court house and the jail. Local police roped off the area, and a bomb squad was summoned from a nearby army base.

After a few minutes of indecision, the bomb squad felt that it was better to blow the top off the box than risk a possible terrorist bomb exploding on its own and causing injury to county employees or destruction to government buildings. The explosion did not fill the air with shrapnel or napalm, but with pieces of shredded newspaper and bits of fried chicken. Then, just a few seconds later, the officers were startled to hear the sounds of frightened mewing coming from the smoking box. A few more moments went by before a badly injured kitten began to crawl out of the package. It looked around at the humans assembled before it, shook its head as if to rid it of echoes of the explosion, then limped beneath a nearby police car.

Deputy Pickett retrieved the mutilated kitten from under the patrol car and instantly decided to adopt it. The name Thunderball seemed very appropriate.

A few minutes later, the deputy had Thunderball at a veterinary clinic where a surgeon removed the cat's left rear leg, patched up his mangled tail, and treated his damaged lungs. The vet told Deputy Pickett that with the resilience that is common among cats, the little fellow was certain to pull through.

*E*ven the most devoted cat lover must concede that for many men and women throughout the world, the cat is associated with the occult, black magic, witchcraft, and the supernatural. And there are those who believe that the unwavering stare of the cat can bring about ill fortune, even cause death. Such an unreasoning fear of cats is known as ailurophobia. King Henry III of England would faint at the sight of a cat. Adolf Hitler had plans to dominate the world with his Third Reich, but the sight of a cat would set him trembling. Napoleon Bonaparte arrogantly snatched the crown of the Holy Roman Emperor from the Pope and conquered nearly all of Europe, but whenever he came upon a cat, he shouted for help. Such a dread of cats may be genetically transmitted, for when Napoleon's brother, Joseph Bonaparte, King of Naples, visited Saratoga Springs in 1825, he complained of sensing a cat's presence just before he fainted. Although his hosts assured His Majesty that there was no such creature on the premises, a persistent search revealed a kitten hiding in a sideboard.

In medieval Europe, black cats were associated with witches and were considered the preferred forms that satanic spirit allies known as "familiars" would assume. The Inquisitors conducting their tribunals against those accused of heresy and witchcraft decreed that all cats were actually demons in disguise, and it is a matter of record that thousands of cats were burned at the stake. It is quite likely that this baseless old ecclesiastical judgment generated the belief that the very sighting of a black cat is an omen of approaching misfortune, and that unhappiness and personal tragedy will soon follow in the wake of the black cat that happens to cross one's path.

There are the many accounts of people encountering the spirits of their deceased cats. Timothy Green Beckley of Inner Light Publishing Company in New York recalled his acquaintance with a ghost cat, the spirit of a favorite pet that apparently survived physical death.

"When I was in my early teens, my family had a very frisky cat named Sweety who had silver-gray hair and beautiful green eyes," Beckley said. "Sweety would hide in all sorts of places in the house and yard. Yet whenever he heard the tinkling of a little bell that we kept near his food and water dish, he would come running at full speed. It was a funny sight to see him trying to scamper across a newly waxed floor, his legs moving rapidly—but his furry body standing still due to the slippery linoleum surface."

Sweety lived to a ripe old age, Beckley recalled, and the family was heartbroken when their lively feline friend finally passed away.

"Many years later," he said, "my father still lived in the same house. Since my sister and her family have a place of their own nearby, they were able to visit Pops more frequently than I was able to do. On several occasions, my sister Bobbyjane swore that she saw Sweety run by her as if he were on his way to eat."

Bobbyjane insisted that she saw the ghost of their favorite pet many times. "In fact, I've almost tripped over Sweety," she said. "He just zips right through the kitchen and then disappears."

Beckley said his sister's testimony has been corroborated. "My nephew Brian has said that he has seen a cat on the stairs leading to his grandmother's apartment. Although he has tried to catch the animal many times, the cat vanishes before his eyes."

Beckley pointed out that Brian is much too young to have seen Sweety in real life. "The cat died many years before he was born. Yet Brian's description of the ghost cat with the silver-gray hair and the beautiful green eyes is so similar as to be nerve-rattling."

*S*eventy-seven-year-old Sarah was terribly distressed when she heard the news about Bucky in September 1982. In tears, she told her husband, Tony, that the beloved cat they had given to friends had been put to death.

"We only gave Bucky up because we thought we were getting too old to care properly for him," she said. "Cissie and Dave had to put Bucky to sleep because the rules of their new apartment house in Miami forbade them to keep pets. Why, oh, why, did they not call us and ask us to take Bucky back?"

Tony was also very upset. They had both loved the tomcat who was nearly fifteen years old. They would gladly have taken back their old feline friend.

They first heard the scratching at their bedroom door about a week after they had learned the sad news of Bucky's execution. The noises became so insistent and so loud that they were both awakened from a sound sleep.

Then, to their astonishment and delight, the elderly couple felt something jump on their bed—and they felt Bucky bouncing across the covers the way he had always moved in life. Tears came to their eyes as they clearly heard affectionate purring fill the room.

"People can call us old fools if they like," Tony said. "But we have a real ghost cat in our home. We know it is Bucky, and he jumps up on our bed nearly every night."

Sarah said she could sometimes feel the invisible cat rubbing up against her legs. "A couple of times we even found cat hairs on the sofa pillow in the living room. That spot used to be Bucky's favorite to sneak up and take a nap."

The most important thing Bucky's ghostly return demonstrates to Sarah and Tony is that their cat did not blame them for his untimely death.

"His coming back lets us know that he still loves us," Tony said, "and that love between animals and humans can reach across the grave."

*M*usic teacher Ruth Wharton, who now resides in La Jolla, California, once owned a white part-Siamese cat named Snooky that she loved very much. "Snooky used to sit outside the door to the house and wait for my music pupils to arrive for a lesson," she said. "By positioning himself right beside the door, he would be certain to be petted by each of the pupils before they entered."

Ruth announced her firm conviction that Snooky was psychic. "He would play and meow to unknown spirit cats that I could not see. Snooky could also tell when my husband would come down the street in his car. Wherever he might be, Snooky would come running as fast as he could to welcome my husband home."

One day after his regular feeding time, Snooky did not return. Ruth said she searched everywhere, calling his name over and over. "But Snooky never came home again."

One month from the day that he disappeared, Snooky's ghostly image appeared in the tree outside Ruth Wharton's window. "I was in the middle of giving a piano lesson when suddenly the pupils turned toward the window—and there on a branch in the tree was Snooky," she said. "It was really my cat, and he went up on the roof and began to meow. I went out on the patio with my pupils and called him, but we never heard him again."

On two other occasions, however, Snooky did return in a similar manner to inform his owner that he was alive and well—just in a different form.

"Such experiences have convinced me that no one ever dies," Ruth Wharton said, "whether you are a human or an animal."

*I*n our book *Dog Miracles*, we told many stories about remarkable dogs that managed to travel astonishing distances to be reunited with their owners. We recounted the incredible adventure of Bobbie the collie, who made his way from Indiana back home to his human family in Oregon, walking 3,000 miles through forests and farmlands, mountains and plains, scorching heat and freezing cold. Nick, a female Alsatian, walked 2,000 miles, from southern Arizona to Washington State to return to her owner. And then there was the nearly unbelievable, but fully documented, account of Joker, a cocker spaniel, who somehow managed to stow away on the very army transport that would bring him to his owner on a secret base in the South Pacific during World War II.

While these marvelous "Lassie Come Home" stories about dogs finding their way back to their owners have become an established element in contemporary animal lore, is it possible that there are accounts of cats that have also managed to traverse great distances to be reunited with their human families? Can it be that felines, almost universally stereotyped as aloof and largely indifferent animals, have courageously braved unfriendly environments, hostile elements, and starvation, sustained only by the abiding and driving motive of being once again in the company of the humans they love? The answer to both questions is a resounding yes!

For many years now, experts in such matters have recognized two basic types of homing journeys that have been successfully undertaken by cats. The first kind occurs when the cat has been given to another owner, when the family moves with the pet to another home, or when the feline has been "catnapped." The challenge presented to the animal in such instances is to find its way back to a familiar home base after beginning the journey in unfamiliar surroundings. If the distance is not terribly far, the cat's keen senses of smell and hearing may help it to negotiate the distance back home in a relatively brief period of time. If the distance is several miles, some investigators of such phenomena believe that cats may have a unique sensitivity to certain of the earth's magnetic forces.

In the summer of 2000, Margaret Bauer of Kearny, New Jersey, accepted the care of a ten-year-old cat named Church that her friend Jim Totin delivered to her home. According to Totin, the Goldberg family, who lived in Eatontown, New Jersey, fifty miles to the south, had decided to give the cat up, and Margaret offered to provide Church with a new home. However, when she returned home from dinner that night around 11:30, the cat was gone.

Back at the Goldbergs' home in Eatontown, Craig Goldberg was shocked when he saw Church sitting on a windowsill, waiting to be let in. The cat was missing whiskers on one side of its head, had a wicked gouge on one cheek, and was limping slightly, but somehow Church had managed to return home—negotiating heavy traffic, a maze of highways and bridges, and a distance of fifty miles—in seven hours. The Goldbergs decided to keep Church, their very own wonder cat.

While the story of Church is remarkable, especially considering that he would have to have averaged a consistent speed of six miles an hour to have accomplished the feat, his dramatic return to his familiar environment may be considered but a mere hop, skip, and a jump in comparison with those cases in which the cat is somehow able to return to its owners after walking hundreds, even thousands, of miles. Accounts of long-distance homing journeys are much more difficult to explain and obviously are accomplished by more than a cat's keen senses and homing instincts.

Such accomplishments as Skittles's dramatic return and other similar stories of remarkable homing journeys recounted below may involve the animal's ability to receive information from levels of awareness presently beyond our human understanding.

Charmin Sampson and her sixteen-year-old son Jason were heartbroken when they had to leave the Wisconsin Dells on their holiday in September 2001 and return home to Hibbing, Minnesota, without their orange tabby Skittles. They had looked everywhere and called his name until they were nearly hoarse, but the cat was nowhere to be found.

On February 4, 2002, 140 days later, Skittles appeared at the Sampsons' home, suffering from severe malnutrition and needing a good, home-cooked meal, but not a great deal the worse for wear. The orange tabby with white paws had managed to find his way across two states and over 350 miles.

Early in 1950, the Stacy W. Woods family moved from Gage, Oklahoma, to Anderson, California, taking with them their yellow cat, Sugar. Then, in June 1951, the Woods's made a decision to return to Gage. Not wishing to uproot Sugar a second time, they reluctantly left their pet with a friend with whom the cat had a good relationship.

In August 1952, fourteen months after they had moved back to Gage, Oklahoma, a cat jumped through the open window of the barn in which they were milking cows, and landed on Mrs. Woods's shoulder. To her astonishment, the animal began to rub itself against her neck in a familiar manner, all the while purring joyously. Taking the cat into both hands for a closer examination, she excitedly announced to her husband that the begrimed, battered, exhausted cat was their very own Sugar.

In an article in the April 1954 issue of *Frontiers: A Magazine of Natural History*, Woods said that he could not believe that a cat could find its way home over a distance of nearly 1,400 miles. Then he remembered that their Sugar had a peculiarly deformed hipbone that had been the result of a broken right rear leg sustained in her kittenhood. When he ran his hand over the cat's flank, he found the familiar deformity. There was no longer any question that Sugar had come home.

In December 1949, Cookie accidentally got shipped 550 miles away from her home in Chicago to Wilber, Nebraska, by Railway Express. Six months later, she had managed to find her way back to her old stomping grounds in the Windy City.

Accounts of cats who find their way home after traversing great distances are truly touching stories of the bond of love that connects feline and human. Apparently in cases of desperation and loneliness, the love connection allows the animal to possess a kind of hypersense that permits it to accomplish homing feats that would seem impossible to achieve.

Now we introduce an even more incredible element into our stories of cats returning to their owners. How can we explain those tales of cats that manage to find their way to beloved owners who are now living in new homes in cities and states far away from the old homestead? Such stories that comprise the second kind of homing journeys, are testimonies to a power of love that can truly soar free of the physical limitations of time and space and permit psychic linkups far beyond the present understanding of our sciences.

Misele could not bear it when Alfonse Mondry, her eighty-two-year-old owner, was removed from his farm and taken to a hospital in Sarrebourg, France, in 1991. What some would term a most remarkable homing instinct, others would call a miracle. Following an infallible guide whose source as yet remains beyond science, Misele set out to visit Mondry in a place to which she had never been. With a determination that would not yield to stone quarries, fields, forests, or busy highways, Misele walked the nine miles to the hospital.

Somehow avoiding the orderlies, doctors, and nurses, Misele located Mondry's room, pushed open the door, and jumped onto his bed. Later that evening, when nurses and doctors found the cat purring contentedly on the old man's lap, they permitted Misele to remain with her master.

In 1953 Chat Beau required four months to hike the nearly 300 miles between his owners' former home in Lafayette, Louisiana, and their new home in Texarkana, Texas.

In 1956 Pooh needed the same amount of time to cover the 200 miles between his human family's former residence in Newnan, Georgia, and their new domicile in Wellford, South Carolina.

Smokey probably rested here and there on his journey from the old homestead in Tulsa, Oklahoma, to the family's new place in Memphis, Tennessee, in 1952; it took him a year to travel those 417 miles.

Tommy required a year and a half to do it, but in 1949 he somehow managed to find his way back home to Seattle from Palo Alto, California—a distance of 850 miles.

When Clementine's human family moved to Denver in 1949, she was left behind on the farm outside Dunkirk,

New York, because she was about to become a mother. Three months later, her coat rough and matted, her paws cracked and worn, her bushy tail dwindled to a rag, she arrived at the front door of the family's new home in Denver.

How the loving and loyal Clementine had managed to negotiate rivers, mountains, and prairies to find her way to a strange house in a city she had never been to remains a mystery.

However, we have even greater mysteries of remarkable homing journeys to share.

In April 1955, when Vivian Allgood, a registered nurse from Sandusky, Ohio, had to move to Orlando, Florida, because of a job change, she was forced to leave behind her beautiful black cat Li-Ping in the care of her sister. Li-Ping moped around his new home for a couple of weeks, but he did not take well to the separation from his beloved Vivian. Then, to the horror of Vivian's sister, he disappeared. In one of the most difficult tasks that she had ever undertaken, she sat down to write Vivian that Li-Ping had grown depressed over her absence and had run away.

One evening in May, about a month after she had moved to Orlando, Vivian Allgood's attention was drawn to a sorry, bedraggled cat limping painfully along in the street. Following one of those strange impulses that seem to come from nowhere, Vivian called out the name "Li-Ping."

Both she and her friend were astonished to see the cat stop dead in its tracks, turn to the sound of her voice, then come limping and stumbling to her just as quickly as its punished condition would allow. Within moments, the cat was in her arms, and Vivian was jubilantly weeping and shouting that it was indeed Li-Ping!

Her friend wanted to know how the cat could possibly be the one that Vivian had left back in Sandusky, Ohio. Many cats look a good deal alike. How could this poor cat, scarred, scratched, and half-skinned, possibly be Li-Ping?

Vivian appraised the shabby condition of the cat in her arms. Large hunks of hair had been torn from its body. Its feet were raw and bleeding, as well they might be after walking 1,586 miles! Vivian knew that it was Li-Ping, who had never meowed, but only made a strange kind of rasping sound. As she lovingly repeated his name amidst welcoming tears of joy, the cat she held so gently made only faint rasping little cries in response.

Each day for a week, Vivian Allgood presented her valiant and courageous cat with his fill of milk and liver until his many wounds and his sore feet began to heal. How Li-Ping could have found his way to a destination far beyond his possible experience or knowledge cannot be explained. The cat had never traveled to Orlando with his owner, and even though he had heard Vivian discussing "Orlando, Florida" with her sister, it is unlikely that the concept would have had any meaning to his cat brain—certainly not enough to have allowed him to find it on a road map and begin the trek to his beloved mistress.

In 1949, Rusty's human family somehow misplaced him while they were visiting Boston from their home in Chicago. After a desperate and tearful search for their missing pet, they gave up the hunt and returned to Chicago, a thousand miles away.

Eighty-three days later, Rusty was back home, scratching on the door to be let in.

Puzzled experts on such strange and unusual matters came to the conclusion that the cat had somehow managed to hitch an occasional ride on a train or a truck in order to traverse such a distance in so few days. But how did Rusty know which trains, trucks, or automobiles would take him in the direction of Chicago, Illinois—and not New Orleans, Louisiana, or Albany, New York?

Tom, a cat belonging to Mr. and Mrs. Charles B. Smith, may hold the feline record for traveling the greatest distance to find its owners' new home—2,500 miles, from St. Petersburg, Florida, to San Gabriel, California.

In their book *The Strange World of Animals and Pets*, Vincent and Margaret Gaddis write that in 1949 when the Smiths decided to move from St. Petersburg, they were concerned about the trauma such a long-distance journey might bring to their cat. Because they had once read that cats often develop an allegiance to a place, rather than to people, they were delighted when they noticed that Tom appeared to have developed a rapport with Robert Hanson, the purchaser of their home. Saddened to part company with Tom, but relieved that he would have a good home, the Smiths made the decision to leave the cat with Hanson.

Convincing themselves that they were acting humanely by not disrupting Tom's normal routine and bringing him all the way across the country, the Smiths bade their feline friend good-bye and left for California. Over and over they told themselves that they had done the right thing in leaving Tom in their former house with his new owner. Two weeks after they had relocated to San Gabriel, however, they were saddened to receive a letter from Hanson informing them that Tom had run away. It seems quite likely at that point that the Smiths regretted their decision to leave their faithful friend behind. Now he had run off and would probably be run over by a car, mangled by dogs, or condemned to wander aimlessly until he starved.

On an August afternoon in 1951, two years and six weeks after their move to California, Mrs. Smith became annoyed by the sound of a cat wailing in the yard. Mr. Smith was given the assignment of chasing the noisy intruder out of their yard, but he was amazed when the cat ran toward him and leaped into his arms. It took him only a moment or two to recognize their old friend Tom.

Mrs. Smith was highly skeptical. They had left a sleek, well-fed cat with Robert Hanson. This scraggly creature was skinny, worn, frazzled, and so weak that it collapsed on the kitchen floor when Mr. Smith brought it into the house. Its fur was bleached by the sun and had come out in handfuls. Its paws were bloody and covered with scabs.

Then the Smiths thought of a sure test to determine whether or not the bedraggled cat was indeed their Tom. They had raised Tom on baby food, and he had developed an unusual fondness for Pablum. Mrs. Smith went to the store and bought some, then put a saucer of it on the floor near the exhausted cat. As battered and beaten though he was, the cat got to his feet and dived into the saucer of baby food right up to his whiskers. The Smiths were now convinced that their old Tom had found his way to their home in San Gabriel, California, from their former residence in St. Petersburg, Florida—way across the continent—to once again become a member of their family circle.

How could it be possible that an ordinary cat such as Tom could find his way to his owners across 2,500 miles of terrain previously unknown to him? Perhaps such accounts as those of Tom and Li-Ping seem so mysterious to us because of our assumption that the mind of these cats is confined to their brain and that they therefore cannot know about things at a great distance from them. Perhaps we should consider the possibility that the mind of these animals is free to scan space and time and gain knowledge of their way home to their owners. If we are able to free ourselves from the species prejudice that maintains that human and animal minds are two distinct and separate entities and accept as true the many accounts that insist that meaningful communication does exist between humans and animals, then we can envision that because the human owners of these pets had knowledge of the way to the new home, such information could be shared with the animal mind. Therefore, in the cases of Tom and Li-Ping, the information of how to travel to the new homes in California or Florida, where they had never been, could have been transmitted telepathically to them by their owners, who did know the way.

In his thought-provoking book, *Recovering the Soul—A Scientific and Spiritual Search*, Dr. Larry Dossey, a physician of internal medicine and a former chief of staff of Medical City Dallas Hospital, suggests a "nonlocal mind" that is not confined to physical points in time and space. Such a modality of mind implies that we and our pets, with our individual minds, are parts of "something larger that we cannot claim as our own private possession."

To fully understand and embrace such a concept requires a genuine humility—a humility that allows us "... to know deeply that consciousness is not the sole possession of an ego; that it is shared by not only other persons, but perhaps by other living things as well. It is humility that allows us to take seriously the possibility that we may be on a similar footing with all of God's creatures."

For those who find the concept of a nonlocal, unbounded One Mind distasteful, Dr. Dossey warns that the reason for this may be that too many people have gradually lost their connection with the natural world. "As a consequence," he says, "our world is now gravely imperiled by our ... lack of sensitivity to the whole. . . . If we wish to preserve our world, we must find our Mind by recovering our connections with the heavens and the Earth . . . [and] to begin once more to talk, with St. Francis, to the creatures."

In June 2001, a black-and-white kitty named Lapis may have bested Tom's homing record of 2,500 miles by about another 500—but she did have some help returning home, so Tom's standard has yet to be beaten.

On April 19, 2001, Jennifer Hill of Boulder, Colorado, spent the day posting flyers all over town with pictures of her beloved Lapis prominently displayed. The little cat had turned up missing, and Jennifer was worried sick. The flyers pleaded with anyone who might have seen

Lapis to please call Jennifer. She even offered a $100 reward for anyone who might have seen Lapis wandering the streets of the city.

Jennifer went many weeks without hearing from anyone who might have seen Lapis. She told the *New York Post* that she remembered crying every day for the little kitty that meant the world to her.

Then, in mid-June, she received a telephone call from someone who had found a cat with her phone number on its collar. But the call did not originate in Boulder. It was a long-distance call from the Yukon— 3,000 miles away.

David Grant told Jennifer that his niece, Susan, had found the cat, but neither of them could believe that a cat from Boulder, Colorado, could possibly be wandering around in the rough north country of the Yukon. The kitty had somehow survived in an area that was heavily populated by coyotes, wolves, and bears.

Jennifer theorized that Lapis had been able to stay alive in such a hostile environment because she was a really good hunter. And though she was small, Jennifer observed, she was really tough.

No one knows exactly how Lapis got from Boulder to the Yukon, but the most likely possibility is that the cat jumped onto a truck that was headed for Canada and managed to stow away somewhere on board without being seen. Once the long-distance trucker arrived at his destination 3,000 miles away, Lapis jumped off the truck and survived in the woods until Susan found her.

Jennifer had Lapis flown home to Boulder, and she reported that once she was back in her familiar surroundings, the little kitten seemed very blasé about her great adventure. She ate a can of tuna, played for a while with her toys, then lay down and took a nap.

*I*n 1986, when the Tialas family moved from Towanda, Pennsylvania, to Forest Lake, Minnesota, Satin, the object of extreme affection from twelve-year-old Sylvia, slipped out of his cage when the family stopped en route at a motel in Illinois. The family searched the adjacent cornfields for three hours before Sylvia's parents convinced their tearful daughter that they needed to resume their drive to Minnesota.

Many times that winter, the Tialases found their daughter sobbing inconsolably over the loss of her beloved Satin. Months later, a former neighbor called them with the remarkable news that Satin had returned to their old house in Towanda. The stalwart cat had accomplished his 800-mile odyssey in about eleven months, but he shunned the neighbors and resisted all their efforts to care for him. While Satin would not accept charity, neither would he leave the premises.

The Tialases made the trip back to Towanda to reclaim their wandering cat and to establish him once again as the king of the household and the master of Sylvia's affections.

*G*eoffrey and Sandra Langrish of Camberley, England, were inveterate cat lovers who had owned Rusty for quite some time, but in the summer of 1986, they didn't know whether they wanted to keep him with the baby on the way. Rusty was a loving cat, but what if he became jealous when Sandra began lavishing attention on the new arrival? One heard so many horrid stories about jealous cats attacking infants, even crouching on their sleeping forms and sucking the breath from their tiny lungs.

And then there were the hygienic considerations.

Picking cat hair out of the baby's formula, for example. Or what if Rusty brought home some feline disease that could be transmitted to human beings? The baby would be susceptible to such alien germs during its first few months.

Rhoda Young, another cat lover who lived in the next town, heard of the Langrishes' plight and offered to take Rusty off their hands. Because they had heard all those homing journey stories about cats that wouldn't stay put in their new homes and somehow found their way back to their former owners, Geoffrey and Sandra put Rusty in a deep basket and covered him with a thick cloth so that he could not possibly see where he was going. They could tell by his nervous movements that he knew something was up, but they managed to keep him covered in the basket until they reached Young's home, sixteen miles away.

Geoffrey and Sandra had barely had time to exchange greetings with Young before Rusty was out the door and gone. It was as if he knew exactly what conspiracy had taken place behind his back, and as soon as the basket was uncovered, he ran between the human beings' legs and disappeared into the village.

Things were tense at the Langrish household for the next few weeks. Nearly every strange sound made them glance toward the cat portal that Geoffrey had set into the kitchen door. They expected to see poor Rusty come staggering into the house, perhaps bloody and beaten, most certainly bedraggled.

And then, of course, there were the attacks of guilt. Rusty had been their faithful cat. How could they have even thought to turn him out?

After a few months with no sign of Rusty, both the hope of his return and the guilt over his disappearance began to fade. The baby arrived, and Sandra and Geoffrey were too busy to think very often of their cat.

And then one day, seven months later, Rusty came walking into the kitchen through his cat portal. He was none the worse for wear, Geoffrey noticed at once. Rusty had actually gained weight, and he looked the picture of feline health. He was, however, very standoffish, Sandra observed. It was as if he were extremely piqued at having been sent away like an unwanted child.

Geoffrey and Sandra wondered where he had been living for the past seven months. He was in much too good condition to have been living paw to mouth in the wild. He must have been boarding with a very attentive cat lover.

As might be expected, Rusty offered no comment that would clear up the mystery. Local animal experts were astounded that he had been able to find his way back home over a route that he had traveled only once before—covered up in a basket.

Rusty seemed only mildly curious about the new

addition to the Langrish family and exhibited not the slightest evidence of jealousy or resentment toward the baby. Geoffrey and Sandra were so moved by the cat's devotion that they decided not to send him away again to Rhoda Young's place or anyone else's. Rusty would have a home with them as long as he wished to remain.

$\mathscr{P}$eggy and Ronald Keaton of Grand Rapids, Michigan, made no secret of the fact that the whole family pampered Princess, their beautiful calico cat. After all, she was gorgeous and had a sweet disposition, so why shouldn't everyone love her and fuss over her? In fact, Princess had been so sheltered that she had been out of the house only once in her three-year life.

In the winter of 1989, Peggy Keaton's mother, who lived in Toledo, Ohio, became very ill, and Peggy left at once with Ronnie, four years old, and Meagan, two, to stay at the woman's side. A few days later, Ronald and their six-year-old daughter, Stacy, climbed into the van with the family dog and Princess to make the drive to Toledo. Although it was only the second time in her life that the cat had been out of the house, Ronald didn't want to leave her home alone.

Somewhere between Grand Rapids and Toledo, Princess vanished. Ronald had made a pit stop at a highway rest area about a hundred miles from home. He was certain Princess was still in the van then because he distinctly remembered having difficulty getting back into the driver's seat because she was up against the window. Stacy got in on the passenger side and moved Princess out of her father's way. Both Ronald and his daughter assumed that the cat moved to the back of the van to find a comfortable spot for a nap.

Ronald did not stop again until he reached his mother-in-law's home in Toledo. It was then, while they were unpacking the van, that they discovered that their pampered Princess had disappeared.

The three children were distraught. What would happen to their beautiful Princess? They imagined frightening scenes.

Ronald and Peggy believed their delicate house pet would stand a slim chance of survival in the cruel world. After all, they kept needlessly reminding one another, Princess had been out of the house only twice in her entire life.

Certain that their beloved cat would perish in very short order, the Keaton family forced themselves to deal with their grief and focus on the needs of Peggy's ailing mother.

Two nights later, a neighbor back in Grand Rapids telephoned the Keatons to inform them that Princess was sitting on the doorstep of their home, impatiently awaiting their return. Somehow their pampered, precious Princess had managed to walk the hundred miles back to Grand Rapids in a remarkable three days and return to the right house. Obviously their darling calico was made of much sterner stuff than any of the Keatons had been ready to acknowledge. The neighbors promised to feed Princess and look after her until the Keatons could come home.

When Ronald and Peggy and the kids returned to Grand Rapids, Princess did not give them a particularly warm welcome. She seemed especially put out with Ronald, Peggy said. "She hissed at him!"

*M*adeleine Martinet of Tannay, France, hoped her cat, Gribouille, would understand, that day in 1987, that it was just that she really could not afford to keep a cat, and her neighbor Jean-Paul Marquart was a kind man who would take good care of Gribouille.

Gribouille, we might imagine, shrugged his furry little shoulders. Such things happened, after all. They were really no one's fault. The fellow Marquart set a good table. And besides, Madeleine would be just across the street and he could look in on her from time to time.

But a month later, Marquart upset Gribouille's perfect world order when he packed up his family, along with Gribouille, and moved to Reutlingen, Germany, more than 600 miles from Tannay and Madeleine. There was little that Gribouille could do en route other than complain loudly, but no one would listen to his side of the matter.

Gribouille barely took time to rest after the long trip to Reutlingen. He ate one last good meal. Then he set out on the return trek to Tannay. It took the determined cat two years to cover the 600 miles that separated him from his mistress and his home in France. When Gribouille reached Madeleine Martinet's doorstep in August 1989, he was starving, bleeding, and nearly blind from a serious infection in both eyes.

Before his ragged body collapsed in her outstretched arms, Gribouille managed to meow. Madeleine understood the emotion in that impassioned meow, for she told her neighbors that Gribouille was home to stay. She would keep the courageous cat forever.

A few doubting Thomases questioned whether it could possibly be the same cat that had left with Marquart and his family in 1987. If this was truly Gribouille, he would have had to negotiate forests, mountains, rivers, and superhighways. In two years he would have been forced to endure freezing winds and rain, snow and hailstorms, scorching sun and lack of food. Not to mention a thousand angry dogs and two thousand jealous cats that would all have tried to eat him or put out his eyes on nearly every step of the journey.

Madeleine Martinet had a cold look and firm words for all those who dared to deny her Gribouille his miraculous accomplishment. "There is no doubt that Gribouille is the same cat that left Germany to return to France," she said. As proof she cited the fact that the old mother cat that had given birth to Gribouille recognized him at once and began to lick his wounds. "Gribouille is fit and well now," Martinet said. "And he will stay a French cat forever."

When our son Bryan and his fiancée Courtnay visited us in May 2002, he reminded Brad of the talking cat that he met in Chicago's Old Town in 1975. From the inquisitive smile on Courtnay's lips, it was apparent that Bryan had related this childhood encounter on previous occasions and had probably met with disbelief. Brad was happy to verify that his older son was not indulging in nostalgic fantasy, but an actual memory of a most remarkable feline named Yama.

In November 1974, Brad was in Chicago to appear

on a number of radio and television shows to promote a recent book. After a late-night talk show, he accompanied some friends to the apartment of Rosemary Stewart, a local television actress, for a small party in his honor. While others helped with refreshments and other arrangements, Brad sat quietly by himself in the front parlor, enjoying a few moments of quiet and relaxation in an exhausting promotion schedule.

His moments of peaceful reflection were interrupted by what he believed to be the sounds of a child calling for its dinner. "Mama," came the voice. "Want nummies. Want nummies now, Mama."

Brad knew from bits and pieces of conversation on the taxi ride over to the actress's apartment in Chicago's Old Town that Rosemary had two sons, but since he had just met her, he had no idea of their ages. Apparently, at least one of the boys was quite young, judging by his word choice and his high-pitched, singsong manner of speaking.

The voice came nearer, just beyond the shadows that led down a long hallway. Brad felt certain that the small boy was watching him from the darkness. The child had to feel like a neglected waif. And it was after midnight. He must be starved, starved and resentful of the men and women who were robbing him of his mother's attention—and his dinner, his "nummies."

Just as Brad was feeling increasingly uncomfortable with his undesired role as an agent of child neglect, the front door of the street-level apartment swung noisily open and two strapping teenaged boys entered. It was quite obviously their home. Did he misunderstand Rosemary? Did she have three children?

Brad introduced himself to the friendly young men, still puzzling over the plaintive voice from the shadows that had cried for food. Then the voice again issued its pitiful request: "Want nummies now! Nummies now, Mama!"

The younger of the two boys walked to the shadowed area just beyond the hall doorway and scooped up a huge gray cat. "Oh, Yama, you big baby," he laughed. "You are always begging for food."

What Brad had believed to be a starving child was in fact a very large and obviously very hungry cat.

By now the actress and the other guests had joined Brad in the front parlor, laden with trays of sandwiches and drinks. "Yama *is* a child," Rosemary smiled. "He is my big baby."

Undaunted, Yama looked up at her and said, "Mama! Want nummies! Now!"

Later, over snacks and drinks, the actress explained that Yama had begun to talk as a kitten, continuing his articulate demands for attention—and especially for food—into adulthood.

Brad noted that some of Yama's words were truncated, such as "want" was really "wan," but the ear of the listener easily supplied certain missing letters to complete the meaning of the feline's requests. Other words, such as "now" and "nummies," seemed to roll off Yama's tongue with little distortion: "now," almost like "me-ow," and "nummies," "numm-eees." And when Yama became impatient, "now," was elongated into "nooooow" with a pronounced accentuation to denote a demand for immediate attention.

In early 1975, Brad returned to Chicago on business and this time brought his family to visit Rosemary and her sons in their apartment. "Hel-row," Yama had greeted the Steiger children to their astonishment and delight. "Hel-row."

When the kids squealed with laughter and asked the big cat what its name was, without missing a beat, he replied. "Yama. Yama. Hel-row."

And then, looking plaintively at Rosemary, Yama emitted the familiar refrain: "Wan' nummies, Mama. Now!"

Some years ago, Brad had an acquaintance, a former police officer, who insisted that a talking cat paid his family a visit when he was a small boy.

As Hank related the story, his family was having dinner one night when they heard a scratching noise at the back door. When his father got up to investigate, he opened the door to find a large black cat looking up at him. Then, to the entire family's astonishment, the strange cat asked, "Come in?"

Hank remembered that his father was always game for any adventure, so he made a sweeping motion with his arm and stepped aside to allow the big cat to enter their home. "My two sisters and I had our mouths hanging open in awe," Hank said. "It was like something out of a fairy tale."

Once admitted into the home, the cat gazed up at Hank's mother and father and pleaded, "Milk? Milk?"

A large saucer of milk was set before the strange visitor, and the entire family stood watching the cat as it hungrily lapped up the milk with its long pink tongue. After the unexpected feline dinner guest had finished the bowl of milk, it briefly allowed Hank and his sisters to pet it as it lay purring contentedly before the kitchen stove.

Then, suddenly, it rose to its feet, as if remembering a prior engagement, and said, "Go now. Out."

Hank's father got up, opened the back door, and allowed the cat to continue on its self-appointed rounds into the night.

"We never saw our strange talking cat again," Hank said, "but, believe me, the incident has come up at every family gathering ever since."

*P*astor Lowell Haugen told us a charming story of a cat named Loretta who became an honorary member of the church congregation that he served in a central Illinois community in 1994.

"I had noticed this skinny, brownish-colored alley cat hanging around the church for a few days before I found her on a rainy October afternoon under the back steps of the parish hall," Pastor Haugen said. "At my approach, she tried to get up, but it was apparent that one of her legs was broken. The poor thing was obviously in a great deal of pain, and it was cowering under the steps, shivering against the cold. Judging from her scruffy appearance, I was certain that she had been struck by a car and had dragged herself into the churchyard to seek shelter."

Among the members of Pastor Haugen's congregation was a young woman who was a veterinarian, so he pulled on a pair of leather gloves and gingerly lifted the cat into a cardboard box for the drive to Dr. Brenda Truesdell's office. "As a child, I had been terribly scratched and bitten by an injured cat that I had wanted to help and had injudiciously picked up," he recalled, "so this time, I was careful not to repeat that painful bit of personal history."

The injured cat offered no resistance, seeming to sense that Pastor Haugen only wished to be of assistance in its time of desperate need.

Once in the veterinarian's office, Dr. Truesdell set the cat's broken leg and put a splint on it. As a precaution against infection, she gave it an injection of antibiotics, then began petting the cat and praising it for its patience and courage.

"You know," she chuckled softly to Pastor Haugen, "there's something about this cat that reminds me of my Aunt Loretta back in Kentucky. Aunt Loretta was always so patient and long-suffering. She went through a lifetime of hurt, yet she never complained or whined to anyone. When she was younger, she even had brown hair, almost this same color. And I'm afraid that, human-wise, she was about as skinny-looking and undernour-ished as this poor cat."

Pastor Haugen laughed at the sudden and unex-pected comparison. "Then Loretta it is," he pronounced. "Since I've rescued this homeless feline and now feel compelled to become a good Samaritan and nurse it back to health, she might as well have a name."

Dr. Truesdell contributed to Loretta's well-being and her rebirth as a distinct entity by waiving her fee for the setting and treating of the broken leg.

Pastor Haugen prepared a small area in his office where Loretta could convalesce. Karla, his wife, donated a worn towel to make the cardboard box in which Loretta slept more comfortable, and Katie, his nine-year-old daughter, offered a small rubber ball for Loretta's recreational requirements. When he and his family opened a can of cat food, Loretta bounded toward the dish, unmindful of her broken leg. It was apparent that the poor cat was extremely hungry and not at all used to such an elegant bill of fare.

As the weeks went by, it became increasingly apparent that Loretta had decided that she had found a permanent home in the church, and Pastor Haugen grew to enjoy having some company in his office when he was preparing a sermon or working on church business.

Even before the splint was removed from her leg,

Loretta had taken to emerging from the pastor's office in the back of the church and walking out to join the congregation just as the Sunday service was about to begin. She would confidently invite herself to join a congenial worshipper on a pew during the opening hymn and watch attentively as Pastor Haugen began the weekly time of praise and thanksgiving to the Creator of all things. During the sermon itself, when all was quiet except for the minister's voice, Loretta would curl up on an accommodating individual's lap and appear to be listening intently, eyes half closed in concentration.

"What was so interesting, and so touching," Pastor Haugen said, "was that once a month during the regular service, our church would hold a special healing ceremony. At that time, those members of the congregation who requested the prayers of their fellow worshippers and the healing energy of Almighty God were able to come forward to the altar and kneel to receive a special blessing.

"The first time that Loretta joined those at the altar who were requesting healing, there were a few giggles and a bit of laughter from the members of the congregation who remained seated. But then it was noticed that Loretta positioned herself right next to a man whom everyone in the church knew was very ill with cancer. The laughter ceased at once when he leaned over to pet Loretta and everyone could see the tears glistening in his eyes. Within moments, he held Loretta tight to his chest and began to sob audibly, as if he was truly sharing his fear and his anguish with a soul that he believed would truly understand his terrible anxieties."

Pastor Haugen went on to tell us of the many members of his church who told him that they were certain that Loretta's presence had brought healing energy to them.

"I am not claiming that that beloved cat had the ability to direct or channel healing toward anyone," he said, "but I am also certain that these men and women, who were feeling so ill and so vulnerable, gained great comfort from the very act of having a cat to stroke as they knelt at the altar. There have been many studies which indicate that petting or stroking a cat relieves tension in humans. There are even programs that bring cats and dogs to hospitals to help cheer the patients."

By the time the Christmas season was upon the congregation, everyone had grown accustomed to the presence of Loretta and most of the members of the church welcomed her joining them for their worship services.

"She's one of God's blessed creatures, too," one elderly gentleman had once said, speaking for himself and the other men and women who stood beside him, nodding their approval of his expressed sentiments.

But then came the event that shocked certain members of the congregation and motivated them to declare that Loretta had worn out her welcome.

The curious cat had made herself a part of every

rehearsal for the annual Sunday school Christmas pageant and had enjoyed being petted and fussed over by every child who was participating in the story of Baby Jesus, the Wise Men, the Shepherds, and the Angels who were heard on high. That year, Katie Haugen, the pastor's daughter, had been chosen to play the part of the Virgin Mary, and she took her role very seriously.

"Katie was the one who always fed Loretta her evening meal," Pastor Haugen said, "and she would practice her lines with great feeling as she mixed the soft and dry foods that were the cat's supper. And just as those times when Loretta was the initial audience for my sermons, she would look upward at Katie with those big shiny eyes and seem to be listening critically to every word."

On the night of the pageant, as soon as Katie and the boy playing Joseph entered the stage area, Loretta left her place beneath the pulpit and joined the two children portraying the Holy Couple. The assembled audience, primarily made up of parents, grandparents, and other family members, had long grown accustomed to the presence of the cat in the church, and there were only light, amused murmurs in response to Loretta's joining the cast. In fact, when Loretta sprawled out next to the cardboard figures of sheep, chicken, and cattle, it almost seemed as if she were becoming a part of the tableau of animals adoring the babe in the manger, a role that was being portrayed by one of Katie's dolls.

"Then, as Katie began speaking her memorized lines, Loretta's ears perked up, as if she recognized some of the words," Pastor Haugen said. "Or maybe Katie had somehow conditioned the cat to expect to be fed when it heard certain words or phrases, for she crossed the stage to Katie's side — and then jumped into the manger."

Loretta disappeared from sight for a moment or two, and then all the children sitting in the front rows awaiting their cue to come onstage to sing "Away in a Manger" began to laugh when the cat's head peeped out of the manger, blinked its eyes, and let out a loud cry, as if prompting the children to sing along.

"Fortunately, it seemed as though Loretta was contented with that small contribution to the performance, and she lay quietly, out of sight, for the rest of the pageant," Pastor Haugen said. "But later, after the last Christmas carol had been sung and the children were being collected by their parents, a number of men and women came up to me to protest Loretta's 'obscene and blasphemous interruption' of the Sunday school program."

Some individuals were incensed that the manger of the Holy Christ Child had been profaned by a cat. Others complained that they had always found the presence of a cat in church during worship services to be disruptive and insisted that it was time for the thing to be thrown out into the streets from whence it had come.

A couple of people even cited the age-old superstitions

about cats being associated with witches and sorcerers and suggested to Pastor Haugen's amazement that Satan, the Prince of Darkness himself, had inspired the cat to leap into the Christ Child's manger and seek to usurp the role of the Prince of Peace.

Later that week there were letters citing the cat goddesses Bast and Sekhmet of the Egyptians, figures of a pagan past, and admonishing the pastor for bringing such a creature into a Christian church. Some letter writers made the association of the cat with the Moon goddess Diana of the witches, and wondered if the pastor had fallen prey to the influence of a witch. Since a good number of the congregation were Scandinavian-Americans, certain of his correspondents reminded him that the symbol for Freya, the love goddess of Northern Europe, was the cat.

Pastor Haugen was stunned by the superstitious fear of cats that such letters and verbal complaints revealed. How a simple act of kindness toward an injured cat had snowballed into accusations of paganism was simply beyond his imagination. How allowing one of God's creatures to enter the church during worship services had been distorted into adoration of ancient cat goddesses boggled his mind.

Pastor Haugen went into his office and closed the door. He needed time to think the dilemma through and to come up with an answer that would satisfy those members of his congregation who had responded so vehemently toward Loretta's intrusion into the manger scene of the Sunday school Christmas program.

As was her custom, Loretta jumped up on his desk, her large eyes gazing into his, her contented purring signaling her love for him. Pastor Haugen reached out to scratch the cat's ears. "How can anyone doubt that the Master Jesus recognized that the divine life force is present in all beings?" he asked aloud.

While it was true that the cat is never mentioned in the canonical books that were selected to be included in the Holy Bible, Pastor Haugen had always felt that the feline had been excluded primarily because the early church patriarchs wished nothing of their faith to have echoes of the religions of ancient Egypt. Or perhaps the unsettled lifestyle of nomadic people made it difficult to become masters of an animal that was not easily domesticated.

Then he remembered a rather obscure book that he had picked up at a used bookstore when he was in seminary, studying to become a pastor. He stood up, ran his right forefinger along the spine of a number of books until he came to *The Gospel of the Holy Twelve*, by Rev. G. J. Ouseley, which had been published in 1923. Admirers of the book claimed that it was a translation of early Christian writings that had been preserved in a Buddhist monastery in Tibet. Whatever its claims of authenticity, there were some provocative passages within its text that seemed to address themselves very well to the problem that was dividing his congregation.

That next Sunday, Pastor Haugen began his sermon by conceding that the text from which he was quoting might be considered apocryphal, but the words had the ring of truth. "I told the story of how Jesus had been passing through a village when he saw a crowd of cruel people tormenting a cat. Jesus commanded them to desist and spoke to them of the law of love and the unity of all life in the family of God. 'As you do in this life to your fellow creatures, so it will be done to you in the life to come,' Jesus warned them."

Allowing that lesson a moment or two to be heard, Pastor Haugen went on to relate a second story of Jesus hearing a young cat cry out for food. After he had fed the cat, he gave it to one his disciples, a widow named Lorenza, who assumed its care. When the villagers saw the great teacher showing such compassion for a cat, they said among themselves that he appeared to consider all creatures his brothers and sisters. Overhearing them, Jesus responded by agreeing that all cats and all animals are the brothers and sisters of humankind. They have the same breath of life in the Eternal. And whosoever cares for and defends such creatures in this life shall be rewarded in the life to come.

"The sermon had the desired effect on those members of the congregation who had slipped back into superstition and begun to see a cat as an agent of evil,"

Pastor Haugen concluded his account. "Loretta was once again accepted as an honorary member of the congregation, free to enter during services and to sit upon any lap that would welcome her. And a few days later, during the Christmas service, several individuals brought Loretta gifts of expensive brands of cat food and a number of cat toys."

Pastor Haugen accepted a call to another church in 1996, and when it was time for his family to leave the congregation, they wished to take Loretta along with them. "It is an interesting thing about cats," he said, "that very often they feel a loyalty and a love for a particular home or place more than they feel those emotions for a particular human family. It was clear to us that Loretta wanted to remain in that church, rather than make the trek to another city in another state with the Haugen family. We said a tearful good-bye. Loretta rubbed up against each of our legs, purred, then walked back into the office that was already occupied by Pastor Jenkins.

"The last we heard about Loretta was from Mrs. Solbeck, a member of the Illinois congregation, around Christmastime, 2000. According to Mrs. Solbeck, Loretta had never missed a church service in all those years since we moved to Oklahoma and had already been outfitted with a colorful neck scarf for that year's Sunday school program."

*V*alera Janssen was jogging with her friend Jessie Witter on the river trail outside of the small town in Idaho where she taught junior high English when a gym bag came bouncing down over the rocks from the bluff above them and landed on the path just a few feet in front of them.

"The paved highway is about forty feet above the trail and runs parallel to the river for several miles," Valera wrote in her report. "Thoughtless litterbugs who toss trash out their car windows have been an occasional annoyance during the three years that I have been running that course, but I figured no one would purposely throw a gym bag out of a car window."

Valera wondered if they should check out the contents for identification so they might return the bag to its owner, but Jessie nixed that benevolent concept immediately and said that she had glanced up at the highway just in time to see someone on the passenger's side of a slow-moving car actually lean out of a window and deliberately toss the bag over the embankment. The bag more than likely contained dirty clothes or some disgusting trash that someone had wanted to get rid of in a hasty, crude—and illegal—manner.

But Valera told us in her written report that she suddenly received a clear image of some living entity crying out for help. "In my mind, I could see some little crumpled, bruised being calling out that it needed rescuing. It was in a cramped dark place, and I knew with a certainty that there was a cat inside that discarded gym bag."

She walked resolutely toward the bag, waved aside Jessie's protests to stay away from some careless litterbug's garbage, and knelt down beside the soiled canvas bag and tugged at the zipper. When she managed to pull it open, she reached in and brought forth a small gray tabby, not more than five or six weeks old.

"It was then very clear to both Jessie and me that some heartless jerk had intended to throw the bag into the river so the little kitty inside would drown," Valera said.

"Instead, the bag had bounced down over the rocks on the hillside and plopped down on the trail in front of us."

Jessie asked Valera how she knew there was a cat inside the gym bag, but all she could answer was that she just "knew." It was as if she had mentally heard the kitten calling for help from some kindhearted person.

"Ever since I was a little girl, I had heard stories about people who claimed that they had practiced telepathy with their cats," Valera said. "And I guess that I had also associated cats with the mysterious. You know, ancient Egyptian pyramids and mummies, witches and black cats flying around on broomsticks, sorcerers and spooky cats with hypnotic eyes. I had never owned a cat. In fact, I had always been a little bit frightened of them, but I really felt that I had entered some kind of telepathic rapport with this kitten and that I was supposed to keep it."

Jessie raised a good point. Perhaps the previous owners, however irresponsible they might be, had tried to get rid of the cat because it had a disease of some sort. Valera promised to have a veterinarian check out the cat's health and give it any shots that it might need.

"I named the cat, 'Bounce,' because of the way that it had come bouncing down the hill that day while Jessie and I were jogging," Valera explained. "The vet said that outside of some bruising, Bounce was in perfect health. He had just made the mistake of being previously associated with inhumane humans."

Valera and Bounce settled in together in her three-room apartment and began adjusting to one another's idiosyncrasies. After they had lived together for about three months, the end of the school year was upon them, and Valera dressed him up with a red bow tie, a little straw hat, and brought him with her to the junior high school on the last day before classes were dismissed for the summer. The girls loved Bounce, cooing and fussing over him, but the boys, struggling with puberty and wishing to appear manly, laughed at his straw hat and bow tie and made feeble jokes about Bounce being a "sissy boy" and a "girly man."

On the drive home that night, Valera stopped for a red light and happened to look down at Bounce on the front seat beside her. "As clearly as if he actually moved his mouth and spoke to me, I mentally heard him say, 'Please, Mama, never do that to me again. It was embarrassing.'"

Valera admitted that perhaps she did not really hear the words quite so precisely articulated, but she insisted that the essence of Bounce's discontent came to her in a rush of emotion and feeling. "I received a number of images of certain of the girls lifting and carrying him around the room and a blur of mind-photographs of boys with their mouths open wide in mocking laughter," she said. "Bounce fixed his eyes upon me until I self-consciously blurted out, 'I'm sorry,' and then, apparently satisfied, he looked back at the street just as the light changed to green."

With summer vacation ahead of her and no real plans or commitments, Valera decided to take advantage of her freedom by trying a number of experiments in telepathy with Bounce. "I knew that if I were patient and persistent, I could become a real 'cat whisperer,'" she said.

After reading a number of books on ESP and paranormal phenomena, Valera understood that having a telepathic exchange with an animal did not mean that she could actually "talk" to them. "If I wished to communicate telepathically with Bounce, I would have to focus on a kind of mental picture of what I wished him to do. Since Bounce didn't communicate with words, we would have to 'talk' through mental images. Although it is impossible to compare animal intelligence levels with human intelligence, I read somewhere that the average cat was somewhere around a two-year-old child in its comprehension level. Therefore, I must use mental images on that level, as if I were attempting to be understood by a two-year-old human child."

After several weeks of tests and experiments, Valera felt that she had achieved great success. "I was able to mentally image certain of Bounce's toys and indicate that I wished him to bring only particular ones to me as I sat in another room apart from him," she said. "After employing a certain technique on a regular basis, we began maintaining a high accuracy of successful attempts."

One night when Jessie came to visit Valera and to see how Bounce was getting along, Valera demonstrated the telepathic linkup that she had established with the cat by conducting an interesting experiment.

"I had Jessie write on a tablet the toy that she wished Bounce to bring to her," Valera said, explaining that Bounce had a lot of toys from which to choose—a rubber mouse, a kitty doll, a little duck, a ball with a bell, and two balls without bells. "I would silently read the toy that she had written on the tablet and project the object mentally to Bounce. Jessie was astonished when Bounce was correct three out of four times."

Not wishing to rationalize Bounce's failure, Valera could not resist pointing out that his "miss" occurred when Jessie wrote "red ball" and he brought the yellow ball. "He did bring a ball," she said. "I tried to focus on the lighter color, but I guess it was difficult to project light or dark."

Valera related a dramatic example of human-feline telepathy that occurred one night as she was dozing in front of the television set. "I guess I assumed that Bounce was sleeping in his bed because it was very late," she said. "Suddenly, I felt a great sense of urgency, and I received an image of Bounce feeling confused and troubled. What was puzzling, was at the same time I received a sense of shame, commingled with fear of discovery."

When she went into the kitchen to investigate, Valera discovered that Bounce had been a naughty boy and had gotten into the garbage pail. He had pushed aside the lid and somehow managed to wedge his head into a can where an irresistible smell had drawn him to what he thought might be a tasty residue to lick—and he had become stuck. She could not help laughing at Bounce's plight as the desperate cat was trying to extricate himself from the tin mask that covered his head.

"That was why I had received the commingled mental images of confusion, fear, and shame," Valera explained. "Although he was frightened because he could not free himself from the can, Bounce was also aware that he had been naughty and had broken a house rule by getting into the garbage. For the next few days, I focused on mental images of being a good cat and not yielding to nasty impulses. I truly believe that anyone can help his or her cat break its bad habits by telepathy."

Valera shared a number of the techniques that she used that summer to develop what she maintains is a greatly perfected "psychic telephone" between Bounce and herself. She believes that if anyone makes a sincere commitment to practice the following exercise on a regular basis, the very least that can occur is an even stronger bond of love growing between you and your cat.

Sit quietly with your cat in a place where you will not be disturbed for at least thirty minutes. Calm yourself

and attempt to clear your mind of all negative and disruptive thoughts.

Take a comfortably deep breath, hold it for the count
of four, exhale slowly.

Now visualize or imagine a large umbrella that
spreads over you and your cat. See the umbrella in a
color that you associate with love, such as pink or blue
or whatever color is special to you. Understand with
deep conviction that this umbrella of love will protect
you and your cat from the influence of all negativity.
It will shelter you with love.

The moment that you have pictured you and your cat
under the umbrella of love, stretch forth your hands over
your cat and visualize the entire area around you filling
with golden light, thus creating around you and your cat
your own spiritual energy field. It is this energy that will
allow you to communicate telepathically.

Now visualize a golden line of love that stretches
from the top of your head and your cat's head and
reaches up to the Divine, the Great Mystery, the Source-
of-All-That-Is. This is your lifeline of love, and it pro-
vides you and your cat with a vital, unending supply of
loving energy. This loveline to the Great Mystery bonds
you and your cat in a complete love relationship. This
connection to the Source-of-All-That-Is permits you to
connect your minds in a harmonious linkup that will
always allow you to communicate on the psychic level.

Visualize the golden line of love connecting you to your cat and the Great Mystery whenever you wish to make mind-to-mind contact.

Herewith is the second exercise that Valera utilized to create a mental connection between Bounce and herself:

Sit quietly with your cat in a place where you will be undisturbed for at least thirty minutes. Calm yourself and clear your mind of all negative and troublesome thoughts.

Take a comfortably deep breath, hold it for the count of three, then exhale slowly. Repeat this procedure once more.

Begin to focus on the thought that you and your cat are one in mind and spirit. Form a mental picture of the two of you in perfect harmony. You will not think of any habits that your cat has that you might consider negative. You must focus only on an ideal image of you and your cat in perfect harmony. You must believe with all your soul essence that you and your cat are now approaching a perfect blending of Oneness.

Once you have fashioned an image of you and your cat melding into a perfect Oneness of mind and spirit, hold that picture fast and slowly inhale and exhale comfortably deep breaths. As you inhale, you are drawing in what the mystics refer to as the *mana* or the *prana* and what martial artists refer to as the *ki* or *chi,* the all-pervasive life force. This is the energy of miracles, and it will permit you to shape the ideal condition of unity and oneness with your cat.

Create and hold fast in your mind the picture of perfect unity and harmony with your cat as you inhale and draw in the *mana, prana, chi*. This energy will give the image of unity and harmony enough strength to hold together while your spirit begins to materialize the picture into physical actuality.

Hold the picture firmly in your mind as you continue to breathe slowly, sending vital energy to your spirit. Be *alive* in the picture. *Feel* it. Keep your mind from all negative thoughts and permit the perfect harmony of complete oneness with your cat to continue to grow.

It is in this state of oneness and harmony, Valera states firmly, that you can develop a telepathic communication and a deeper love between you and your cat.

Doug Kessler said that he was aware of recent research that claims that domestic cats actually learned how to create those pleasant mewing sounds in order to ensure that they would be taken in and cared for by humans. Angry-sounding, growling felines—or those with irritating high-pitched screeches—are likely to be left outside in the cold. Indeed, 7,000 years ago in ancient Egypt, when the ancestors of today's domesticated cats were trading their rodent-snatching skills for shelter and the occasional bowl of milk, it was quite likely the more pleasant-sounding cats that were selected to guard the granaries.

In spite of acknowledging such research, Doug, who lives in a suburb of Sacramento, said that he was attracted to his cat Frog because he doesn't meow at all.

"Ever since I got him as a small kitten in the spring of 1994, Frog has only made a croaking, raspy sound," Doug said. "When he gets mad or agitated, he makes a really weird kind of prolonged hissing noise. The day that I first brought him home, my sister said that Frog sounded like the creep who used to call her and make disgusting whispering noises before he hung up."

Frog grew to attain considerable bulk, right around seventeen pounds. "Friends were always telling me that I needed to put him on a diet, but I figured he was just a big guy. And the bigger he grew, the louder his strange croaking hiss became," Doug said. "I know that most cats make that growling, hissing, spitting noise when they're angry with each other or trying to scare away the neighbor's dog, but Frog would blast off like a tire going flat or someone letting the air out of a huge balloon. If he ever had a territorial thing with another male cat, just his loud hiss alone would intimidate the other tom to the point where it would usually turn tail and run away."

Doug was twenty-three when he adopted Frog and brought him into the house that he shared with his parents and younger sister Lorraine. "I was still in law school, living at home with Mom and Dad to cut down on expenses," he said. "In 1996, Lorraine got married and moved to La Jolla. In July 1999, Mom and Dad got tired of the work involved in caring for a house and yard and decided to move into a condo. Frog and I stayed in the house to maintain the place until we decided to put the place on the market. My wife, Amber (at that time, my fiancée), and I were planning to be married in April 2000, and we were talking about moving to Portland (which we did in 2001)."

One Friday night in November 1999, Doug had gone to bed with a splitting headache. "It had been the mother of all wicked weeks at the law office, and I was exhausted," he continued his story. "As a junior partner, it seemed that all the tedious, busy work was always dumped on me. I must admit that I was feeling a bit the martyr, so I had a couple of extra-strong vodkas, and I fell asleep as soon as my head hit the pillow."

At about three o'clock in the morning, Doug was awakened by a man's voice shouting for help.

"At first I thought I was dreaming," he said, "so I just rolled over and tried to fall back asleep. Then I sat bolt upright when I realized that the voice was coming from downstairs in the kitchen."

Doug quickly called 911, whispered his address, and explained that someone had broken into his home. Then he grabbed a flashlight from a drawer in the nightstand, a golf club from the bag in the closet, and summoned his courage to go downstairs to investigate.

"As my senses cleared, I realized that the man was shouting, 'Help me! I surrender! Please help me!'" Doug said. "I walked cautiously down the stairs, having absolutely no idea what might confront me."

Doug heard Frog croaking and hissing up a storm in the kitchen, and when he directed the flashlight beam into the room, he was astonished to see a man, a complete stranger to him, standing on the kitchen counter. Doug also saw that a pane of glass had been broken in the outside kitchen door to allow the intruder to reach in, turn the knob, and gain illegal entry to the Kessler household. It was then immediately apparent that the stranger in his kitchen was a burglar who had intended to rob him. A burglar who had been heard entering the house by Frog the guard cat, who had immediately gone into attack mode.

When the man noticed the flashlight beam, he turned in Doug's direction and said, "Okay, man, woman, or whoever, I surrender. Just call this thing off."

Doug clicked on the light, and the man blinked and cursed at the sudden brightness. "Where is it?" the burglar squinted, looking around on the floor. "Where is that slimy, slithering, bone-crushing monster?"

"What monster?" Doug wanted to know as Frog crossed the kitchen to stand at his side. He concluded that the intruder must be high on some psychedelic drug.

Their dialogue was interrupted by two police officers who had responded to the 911 call and were pushing open the outside kitchen door with handcuffs in hand.

It wasn't until the police had booked and questioned the burglar that Doug heard the full story. The officer who called Doug the next afternoon was laughing as he told him that when the burglar entered the darkened kitchen and heard the loud hissing sound that Frog made, he thought that it was a large snake slithering toward him.

"According to the police officer, one of the burglar's buddies had broken into a home a couple of weeks before that had a menagerie of reptilian pets, including a very large Burmese python," Doug said. "The burglar who was met by Frog in the kitchen had a near-phobic fear of snakes, and when Frog let loose with his special variety of loud, sustained hissing, the terrified man thought that he had stumbled into another California household that loved large, crawling reptiles. Since it was completely dark in the kitchen, there was no way that he could see that there was no monstrous, bone-crushing python at his feet — just a big overweight cat with a weird hissing croak instead of a pleasant meow."

*S*hortly after she had given
birth to four kittens,
Missy was struck and
killed by an automobile as she followed her owner,
Ginny Sutton, to the mailbox at the end of their lane.
Although the Sutton family, who live in a rural area of
West Virginia, had assumed that they would have to take
over the care and feeding of Missy's kittens, they were
surprised when Rosie, their two-year-old terrier, stepped
in to fill the mother role.

Rosie and Missy had been extremely close and the
best of friends. "The age-old hostility and rivalry

between dog and cat simply didn't exist for those two," Ginny said. "They had played together, slept together, even ate together, ever since we got Rosie as a pup."

Rosie had never had pups of her own, but as soon as Missy gave birth to her four kittens, she was right there at her friend's side, licking the kittens and watching over them as if they were her own. After Missy was killed, the Sutton family—Ginny, her husband, Craig, and their thirteen-year-old daughter, Tracy—were touched when Rosie lay down with the kittens in a posture as if she expected to nurse them. At first it appeared as though the kittens were just cuddling up to Rosie for comfort and warmth, but when Ginny looked closer, she was astonished to see that Rosie was producing milk.

"We thought such a thing was impossible," Ginny said. "How could a female dog that had never been pregnant produce milk on demand for a bunch of kittens?"

But their veterinarian, Clyde Martindale, informed them that a female dog can produce milk if its body is signaled to do so. While it may have been better for such small kittens to have received milk from their own mother, Rosie provided a marvelous alternative source.

Such a seemingly phenomenal occurrence is not without precedent. Some years ago, the Briggs family of Milwaukee, Wisconsin, reported that their female German shepherd had assumed the mother role when it adopted a stray kitten. In this instance as well, the dog not only began licking and grooming the tiny stranger, but within a day or so began producing milk and feeding it.

The maternal instinct is just as strong among cats. In their book, *The Mythology of Cats*, Gerald and Loretta Hausman tell of three terrier puppies that had been orphaned and motherless until a female cat in the neighborhood that had just had kittens took the pups into her fold and fed both her own and the orphaned ones without prejudice.

Michael Joseph recounts two such amazing instances of the feline maternal instinct in his book *Cat's Company*. Blackie, a half-Persian, adopted a motherless baby turkey, and even brought mice for the tiny fowl to eat. Puzzled why the turkey's small beak couldn't somehow swallow the mice, Blackie continued bringing different food items to nourish the orphaned chick. According to the farmers who witnessed the strange pairing, Blackie continued to watch over the turkey even after it mated and had a family of its own.

Joseph tells an even more remarkable story in the instance of a barnyard cat in Beaumont, Jersey, England, that adopted a baby mouse. The cat had eaten the little rodent's mother with great relish, but then, instead of bringing the helpless mouse back to her own kittens to be used as a live demonstration of hunting skills, the cat had pity on the tiny thing and brought it back into her fold and began nurturing it together with her own kittens. Several weeks later, the mouse, having survived the attentions of his feline siblings, wandered off into the nearby woods. Then, as astonishing as it may seem, the mother cat, distraught at losing one of her babies, yowled mournfully into the night until the wayward mouse returned home to the strangest nest that a rodent has ever occupied.

On the other end of the scale of the "most amazing accounts of the feline maternal instinct" must be the story carried in *The Moscow Times* in November 2001 about the Siamese cat that adopted two wolf cubs in a Russian zoo. According to Rosa Kaipberdyeva, a spokesperson for the Novosibirsk Zoo in western Siberia, Musya, the Siamese, had become a true mother to the cubs.

The wolf cubs, whose natural mother failed to produce adequate milk to nurse them, were only a few days old when they were brought to Musya, who had one kitten of her own. Because they were so young, the eyes of the cubs were not yet opened, so they couldn't see their adoptive mother. Two weeks later, when their eyes opened, they couldn't have cared less that the generous mother who had been providing such good milk was a Siamese cat. Zoo officials removed the cubs from Musya and her kitten when they were old enough to eat meat.

ark Andrews of Indianapolis, Indiana, shared a fascinating account of how he received assurance that the spirits of his best friend James Walter Fortmeyer and James's cat Sam had been reunited and were at peace on the Other Side.

Mark writes that he and his friend shared a great interest in all things metaphysical and spent many years in the pursuit of truth, often joining various study groups together. "Jim was always there when I needed any kind of help," Mark said, "and he never asked anything in return. There came a time in 1988 when my bills had exceeded my income, and I found it necessary to accept his generosity and move into his house for a time."

Another resident of the house was Jim's cat, Sam. "He had a very mellow personality and displayed many of the 'other worldly' mannerisms that come to mind when we consider the mysterious nature of domesticated felines," Mark said. "He was a very special cat, and Sam and Jim were very close. Sam was already twelve years old when I moved in with Jim, and soon thereafter, the cat died of old age."

Jim was naturally distressed at the loss of his beloved cat and provided Sam with an honorable burial.

In the summer of 1990, Jim moved to nearby Bloomington, Indiana. Mark had recovered enough financially to afford a duplex apartment in Indianapolis. Shortly after moving to Bloomington, Jim began to succumb to a fatal illness.

"He and I had plenty of time to make preparations for his passing," Mark said, "and we made a pact that Jim would give me a sign once he had made a safe transition to the other side."

Jim died four years later on September 8, 1994, and his friends honored his request to hold an official memorial service for him on September 11 in Columbus, Indiana, where the majority of his family lived. After the memorial, a number of Jim's closer friends, who shared his metaphysical interests, gathered at Mark's apartment to share fond memories of Jim and to await with hopeful anticipation the sign that he had promised to send from the Other Side.

Mark had moved into the apartment on September l. The landlord had seen to it that everything was fresh and clean. Even the walls were newly painted.

"When I arrived home from Jim's funeral, I busied myself with all of the last-minute preparations for the group of friends who would be coming over to await a sign from Jim," Mark said.

"When I went to the pantry, I was just a little stunned—and elated—to find *cat's paw prints that had apparently been walked up the side of the wall.* They weren't there when I left for Jim's funeral. I knew at once that those paw prints had to belong to Jim's faithful cat, Sam. It was Jim's way of telling all of us—just as he had promised—that he was conscious and well and had been reunited with Sam. Jim saw to it that our memorial service in his honor was indeed one of great joy and celebration!"

This is a story that could have had a very grim and sad ending if it hadn't been for a protective Siamese named Sybil. Nine-year-old Pam Price, who lives with her parents in a suburb of Detroit, was pushing a doll carriage briskly ahead of her as she walked to the home of a friend to play house with their dolls. Unaware that ever since she left her home she had been followed by a stranger in a dark four-door automobile, Pam was taken completely by surprise when he suddenly slammed on the brakes, dashed out of the car, and grabbed her.

The frightened girl didn't even have time to scream before the brute had slapped a thick band of tape over her mouth and was trying to drag her into the car. But in her struggle to get free, Pam did manage to reach in the little carriage and pull her "doll" out from under the blanket where it had been taking a nap on the way to the friend's house. Pam's "dolly" was a twenty-pound, three-year-old Siamese cat named Sybil, who loved her little mistress enough to die for her.

Fortunately, no such sacrifice was required of Sybil that afternoon in April 2001. The big Siamese lunged forward and slashed the man viciously on the arm that was attempting to keep Pam from struggling as he dragged her toward his car. Startled by the attack that had seemed to come from nowhere, the kidnapper cried out, screaming in pain and cursing at the snarling cat. He let Pam fall to the sidewalk while he turned his attention toward Sybil.

According to witnesses in the neighborhood, the man appeared to be reaching for a weapon in his trouser pocket, but he never had a chance to use the knife or whatever he hoped would silence the angry Siamese. Sybil leaped for the man's chest, sunk her claws into his shirt, and savagely bit at his throat. Howling in pain, he grabbed Sybil with both hands and threw her to the ground.

By this time, several residents had called the police, and two men and four women had left their homes armed with baseball bats, golf clubs, and walking sticks

to encircle the would-be child molester and see to it that he stayed put until the squad car arrived. Someone in the neighborhood who recognized Pam called her mother, Regina Price, and she was there to comfort her daughter as soon as she could run the two blocks to the place where the incident had occurred. She was horrified to learn that her nine-year-old Pam who had just left home to walk a few blocks to her friend's house had narrowly escaped being assaulted and kidnapped.

After the man had been handcuffed and placed in a squad car by two police officers, Pam began to cry and become frightened. It was as if the terrible assault had occurred in such a blur of motion that she was responding in a kind of aftershock. Later, they learned that the forty-year-old assailant was a known sex offender. He was later charged with kidnapping and criminal sexual assault, which is punishable by up to life in prison.

Sybil was the heroine of the hour. She had been dashed to the ground and momentarily stunned by the monster who had grabbed Pam, but she quickly regained her equilibrium and crawled back into the little carriage, ready once again to pretend to be a doll and play house with the girls. Later that night, Sybil received an extra can of her favorite cat food, seemingly oblivious to the great act of courage that she had displayed while defending her beloved mistress.

$\mathcal{A}$nn Goranson told us of the time in February 1998 when she was recuperating from a skiing mishap at the small home that she owned near Winterpark, Colorado. "I had pulled a number of muscles in my back and shoulders, and I was in severe pain," she said. "The only company that I could bear at that time was my old buddy Lorenzo, a cat of patchwork ancestry that had been my faithful companion for nearly ten years."

Ann had declared as diplomatically as she could that if everyone would leave her alone so that she might do some cautious stretching exercises and get some much-needed bed rest, she would probably be able to be back at work at the sporting goods store in a few days.

"I had politely asked all my friends and family to stay away and not even call me so that I would have to get up to answer the phone," she explained. "Everyone abided by my wishes except my boyfriend Wayne, who insisted on calling every six hours or so to see if there was anything that he could get for me."

On the third night of her imposed isolation, Ann was in so much agony that she decided to take both a prescribed pain pill and an over-the-counter sleeping aid in order to get some much-needed rest.

"The extreme discomfort of my mishap on the slopes had been preventing me from getting any deep sleep at night," she said, "so I clicked on the answering machine to intercept any of Wayne's checking-in-on-me calls, turned off the volume on the bell, and lay back on the leather sofa in front of the fireplace, heavily sedated and already falling asleep."

Ann fell into such a deep sleep that several hours later she didn't hear her smoke alarm go off.

"I remember that I was dreaming that I was walking in a zoo with my six-year-old niece, Susan, and my eight-year-old nephew, Jeff," Ann said. "Every time that I would try to hold their hands so we could stay together as we moved among the other visitors at the zoo, they would bite my hands. I scolded them for being such awful little monsters, and this time they bit me on my legs."

That was when Ann awoke to discover that it was not Susan and Jeff who were biting her hands and her legs, but Lorenzo. He had bitten her several times on her wrists, hands, and legs, hard enough to draw blood.

"I screamed at him, thinking that he had gone crazy," Ann said. "He had never done such a thing before. He had never even nipped at me when we were playing or when I gave him a bath. I started to reach for a shoe to smack him. And then I began to cough from the smoke that was filling my little home. Horrified, I saw that flames were burning away at the carpet and moving toward me."

Ann could hardly breathe as she staggered to the door. The sharp intake of winter air helped clear her senses, and she began to scream for one of her neighbors to call the fire department.

"Although the firemen did their best, my cute little house was pretty much destroyed," she said. "I did manage to salvage a few personal possessions that weren't too badly fire damaged, but my comfortable little home was burned beyond repair. Of course, the important thing was that Lorenzo and I had escaped being burned to ashes along with most of my belongings. For a while, I forgot the awful pain in my back and shoulders as I stood outside in the cold, sobbing, holding Lorenzo, watching the firemen try to save what they could of my home. When Wayne and some of my friends arrived, I told them how Lorenzo had saved my life."

Later, when the fire inspector offered a reconstruction of how the fire had begun, he suggested that some large sparks had managed to fly over the fireplace screen and land amidst the balls of multicolored yarn that Ann had left on the floor.

"I was in the process of knitting a ski sweater for Wayne," she said, "and I left the needles and yarn in a mess when I started getting groggy. The fire inspector theorized that the soft balls of yarn had slowly smoldered into flame, and the fire had eventually begun spreading across the thick carpeting and moving toward the sofa where I lay sleeping. I was so sedated that I didn't hear the smoke alarm blasting a warning. It was obvious from the wounds on my hands, wrists, and legs that Lorenzo had bitten me many times in a desperate attempt to awaken me."

A few weeks later when Ann had found an apartment and was entertaining some friends, she recounted the story of how her heroic cat had saved her life. While most of her friends were aware of Lorenzo's heroism and the place that the cat held in Ann's heart, a skeptic among them began to question Lorenzo's heroism. He laughingly suggested that the cat had simply responded to its own instinct for survival. According to his interpretation of the near-fatal fire, Lorenzo had panicked when the smoke was billowing up around him and he felt the heat of the flames rising in the small house.

"Your little hero Lorenzo had probably gone berserk with fear and was biting you only so you would wake up and open the door so he could escape the fire and flee into the night and save his own fuzzy tail." He smiled, pleased with his clever and rational analysis of a feline's alleged act of courage.

Ann quickly demolished the cynic's theory. "I told him that my former home had a kitty door," she said. "If Lorenzo had panicked at the onset of the fire and had thought only of himself, he could have made a dash out of the kitty door any time he wanted to and left me to burn to death. Instead, Lorenzo chose to stay behind with me, trying his valiant best to awaken me before the fire or the smoke killed me. Thank God, he succeeded, for his sake as well as mine — for I know that he never would have left me to die alone."

A newspaper reporter who wishes to be known only as Colleen was feeling very anxious as she left the press room that evening after working late into the night. A vicious night stalker had been committing violent assaults on young women in the midsized city in the northwest where she lived and worked, and when Colleen saw that a number of streetlights were out on the block ahead of her, she wished that she would have called a cab instead of deciding to walk home.

The monster's method of unleashing his hatred of

women was to grab them while they were walking down a dark street, drag them into a car, and beat them mercilessly. In the past month, seven victims had fallen prey to the brutish assailant. Four of the women were still in the hospital—one in critical condition.

In her column commenting on city life in that afternoon's edition of the newspaper, Colleen had called the serial attacker of women a coward and a sick individual in need of help. That night before she left the office, she had received two telephone calls that consisted only of someone uttering an obscenity in a deep, rasping voice. She couldn't help wondering if she might be the next targeted victim of the night stalker.

"I was just a little more than a block from my apartment building when I was surprised to see my big old Maine coon cat Rhoda ahead of me on the sidewalk," Colleen said. "There was no way that Rhoda could have gotten out of the apartment unless my roommate, Abby, had neglected to shut the door behind her. I am very strict about observing both animals' and people's rights, so Rhoda wears a little belled collar at all times and she is on a leash when I take her for a walk. Abby had never taken Rhoda for a walk, and she was not particularly fond of her."

Puzzled by her pet's unexpected appearance, Colleen called for Rhoda to come to her, but the cat arched its back and began to emit the loudest, eeriest yowling that the reporter had ever heard. "Rhoda just sat there in the middle of the sidewalk, literally sounding like a lost soul wailing in hell," Colleen said.

Colleen was kneeling to attempt to comfort Rhoda when she was suddenly grabbed from behind. "You can imagine the thoughts that swirled through my consciousness," Colleen said. "I could feel strong arms squeeze me around my abdomen. I was lifted off my feet, and my assailant was beginning to drag me toward an alley."

That was when the big Maine coon cat—still emitting the fearsome yowling—leaped into action, clawing the man's hands and causing him to release Colleen. "Once I could breathe again, I joined Rhoda's screeching by screaming at the top of my lungs," Colleen said. "I screamed for help and waved at a passing motorist, who slowed down to see what the fuss was all about. At that point, my attacker set off running into the darkness of the alley."

Colleen never knew if the brute who attacked her a block from her apartment building was the same assailant who had so viciously beaten the seven young women, for the savage assaults ceased as suddenly as they had begun. While police officers puzzled over the mystery of the night stalker's disappearance, and the women of the city gave special thanks that he appeared to have vanished into the same hellish pit from which he had come, Colleen tried to solve the enigma of how

Rhoda had managed to appear on the sidewalk that night just in time to save her from what surely would have been terrible injury or possible death.

"Abby insisted that when she left the apartment on her date that night, Rhoda was sleeping soundly in her basket," Colleen said. "When Rhoda and I arrived home that night after I had given a statement and description of my attacker to the police, I found the door securely locked. I carefully checked all the windows, and none of them were open even a crack. Besides, we lived on the sixth floor and that would have been quite a drop even for a cat.

"Somehow, in a way that I will never be able to understand, Rhoda must have sensed the danger that I was about to face and *somehow*, she was able to appear physically on the street to rescue me," Colleen said, concluding her account. "I keep thinking that there is some rational explanation as to how Rhoda could have accomplished what appears to be an impossible feat—that is, leaving a locked apartment and materializing in front of me on the sidewalk. Then again, maybe even a hard-headed journalist like me has to admit that on occasion a miracle occurs that has no logical explanation."

*D*r. Ingrid Sherman shared a story of the power and the ingenuity of a female cat's maternal instincts. "My family and I were spending the summer at our cottage in the mountains," she writes. "Rather than leave our beloved pets behind in the care of others, we took our dog, Happy, and our two cats, Fluffy and Tiger, with us."

Tiger was pregnant, and Ingrid was certain she brought along a box for the female cat's birth-giving. "Tiger was familiar with the procedure," Ingrid explained, "for when she had given birth in the past,

I had designated the box as her 'nursery' in a corner of the garage back in our city home."

When it appeared that Tiger was about to give birth, Ingrid placed the box under a tree in back of their mountain cottage. However, the next morning when she went to investigate, she found that Tiger had forsaken the birthing box in favor of a hole that she had dug under a bush.

"How clever," Ingrid observed. "Tiger had assessed the cooler temperature in the mountains and concluded that the night cold would be harmful to her two new kittens. She had dug a hole under a thick bush and raked in a thick layer of leaves to serve as a protective covering while she brought her new babies into the world. Both of the kittens were all black with no other markings."

That night, there was a heavy rainstorm in the mountains, and Ingrid feared that Tiger and the kittens would be soaked and chilled by the downpour. "I lay there worrying about Tiger and her two kittens," she said, "but I didn't want to go out into the night and disturb them. Some mother cats can become very upset in those early hours after they have given birth, and Tiger was one who prized her privacy at this time."

Early the next morning when she went outside to investigate the state of Tiger and the kittens, Ingrid was astonished to find that Tiger had carried the kittens to the shelter of the cottage's covered porch and had placed them in two old shoes that had been left there. "The kittens were resting comfortably, safe from the rain, in those two old shoes," Ingrid said. "How cleverly and admirably a cat's maternal instinct can display itself."

As a postscript to her story, Dr. Ingrid Sherman states that Tiger's kittens grew into healthy specimens. "A lady bought those two black cats from me, because she wanted to add them to her all-black living room with black carpeting and black couches so that they might keep company with her black dog and black bird."

*I*t didn't matter one bit to Bonnie, a six-year-old tortoiseshell, that the thieves were attempting to steal bags of *dog* food. All that mattered to her was that someone had broken into her master's warehouse in Derbyshire, United Kingdom, with the intent of making off with property that did not belong to them.

When Mike Powell opened the door to his warehouse that Sunday morning in May 2001, he was startled to discover the place in disarray. A quick inventory told him that a few bags of dog food were missing, but there were streaks of human blood and bits of cat fur that remained to give testimony to a scene of violent hand-to-claw combat.

Bonnie approached Powell, looking somewhat sore and slightly the worse for wear, but there was a gleam of triumph in her eyes. Although Bonnie had obviously been battered around a bit and had tufts of fur torn from her body, it was also apparent that she had given a great deal worse than she had gotten.

Judging from their detection of two distinct types of human blood and a variety of other physical evidence left on the scene, the police officers' reconstruction of the previous evening's battle royale indicated that at least two thieves had backed a truck up to the warehouse with the intent of stealing several tons of dog food. They had loaded only a few bags into their truck when Bonnie came upon them during her nightly surveillance of the premises and attacked them with fangs and claws set for maximum fury.

Although the burglars had fought back and battered the tough cat quite a bit in their defense, Bonnie had splattered enough of their blood around the place and had caused them to leave behind several other items of evidence that the police were confident would lead to the intruders' apprehension.

Powell praised Bonnie, the guard cat that had saved him a fortune in dog food, and said that she was easily worth her weight in gold.

*M*any cat owners believe that it is important to put a great deal of time and thought into the process of selecting the proper name for their cat. According to the beliefs of some individuals, choosing just the right name for one's cat is every bit as essential as choosing the best possible name for one's child. A name, they say with conviction, is concerned with sound, a direct manifestation of vibration, and influences both the inner and outer expression of a human being or an animal. What name you give your cat, they insist, will have a great deal to do with its behavior and the development of its personality. As T. S. Eliot wrote in *Old Possum's Book of Practical Cats*, which served as the inspiration for the long-running musical *Cats*, every cat needs a name that's "particular . . . peculiar . . . and . . . dignified, else how can he keep his tail up perpendicular. . . ."

Because the domesticated cat traces its origins to ancient Egypt, many cat owners have selected Egyptian-sounding names. Several of those who fancy Angora cats have done some research into Persia or Turkey, the area where that breed originated, and found names derivative of those cultures. Siamese cats have prompted names with an Asian lilt, and long-haired Siberians are blessed with Russian-sounding monikers.

Some cat owners advise waiting a few days after you have acquired your new feline to see if it will reveal any unique personality traits that might suggest a name. Others have expressed their opinion that the name that you give your pet will do a great deal to influence its temperament and personality.

It seems good advice to choose a name that is short, for it has been demonstrated that both cats and dogs respond more quickly to names of one or two syllables. Although some folks enjoy giving their cats lengthy and impressive titles, such as "Montgomery Wellington III," the name will soon become abbreviated to "Monty" or something much briefer and more familiar. It is inevitable that any long name will soon be shortened. The test of an appropriate name usually comes when one is calling the cat. As in the previous example, the vast majority of people would soon tire of shouting out across the neighborhood, "Come home, Montgomery Wellington III!"

And speaking of summoning your cat, it is also wise to choose a name that is both simple and easy to call out—and one that sounds respectful and cheerful. How much more pleasant to call out "Elsie" into the neighborhood after dark than "Killer" or "Tuna Breath."

*J*t wasn't like Tigger, her big tabby cat, not to return home to Caridwin Jones, a sixty-two-year-old nurse, when evening was beginning to darken the streets of Devon, England. It was a cold night in February 2002 when Tigger failed to walk into the warm kitchen and the loving family of Mrs. Jones, her three other cats, and a parakeet.

After a few days had passed, she feared that someone had catnapped Tigger, wanting to keep him for their own. She went to the local office of the Royal Society for the Prevention of Cruelty to Animals, asking for assistance in reclaiming her beloved cat. In addition, she put up posters in the local pet shops in Devon, asking that Tigger be returned to his rightful home, and she broadcast an appeal through the radio stations that the big tabby be allowed to come back to her.

Weeks later, with absolutely no response from any guilty catnappers or any person who had spotted her cat, Caridwin Jones was forced to come to the grim conclusion that someone was keeping Tigger, unmoved by her appeals for his return, or that he had somehow been killed and his body lay unnoticed in some remote area.

Although she disliked even thinking of such a possibility, Mrs. Jones knew that a landslide caused by a collapsing wall had occurred near her home about the time that Tigger had disappeared. It was dreadful to think that her dear cat could have been covered by tons of dirt and rubble.

Then around midnight, a month after Tigger's disappearance, Derek Hope and Wendy Howard heard a strange sound coming from a partially open window in their home that had been covered by the landslide. To their astonishment, a skinny, bedraggled cat was slowly digging its way through the window and into their home.

Once it had managed to wiggle its way through the rubble, it was apparent that it was so weak and disoriented that it could not walk. Wendy took the cat tenderly in her arms and began to comfort it. She knew that the poor starved cat, nothing but skin and bone, had to be Caridwin Jones's missing Tigger.

As Derek Hope recreated the scenario of Tigger's dreadful ordeal, he suggested that the tabby was probably nearly home when the wall collapsed and buried him under tons of dirt and rubble. Somehow, he found himself in an air pocket that kept him from being crushed, but it had also trapped him—and it had taken him a month to slowly dig and wiggle his way through the rubble. Tigger had managed to stay alive by licking moisture that had condensed on the pieces of brick and mortar mixed in with the dirt, but he had emerged a mere skeleton of his former self.

Derek Hope theorized that even though the poor cat was starving, the more weight he lost, the easier it became to drag his thinning body through the debris. Grateful to have Tigger back in her family, Caridwin Jones observed that her tabby had to have used up at least four of his nine lives by surviving under a landslide for a month.

*C*ats probably dislike having needles stuck in them just as much as their human owners do, but a few years ago we heard about a cat in Norman, Oklahoma, named Ramone, who had become a regular blood donor in the office of veterinarian Dr. John Otto for cats needing transfusions.

It was Dr. Otto's wife, Patti, who found Ramone when she went to an animal shelter in search of her own missing cat. She didn't locate her pet, but she did spot Ramone in a cage with over a dozen other kittens. While the other tiny kittens were meowing and climbing over one another,

Ramone sat there calmly, seeming to look directly at her.

Patti Otto said that her heart went out to the kitten, especially when she learned that the entire batch of cats were scheduled to be put to sleep in another day or two. She took Ramone home with her, and he was a member of their family for seven years before he became a blood donor in Dr. Otto's clinic.

One day a cat suffering from a disease caused by fleabites was brought into the clinic. Dr. Otto knew at once that flea anemia would soon kill his new patient unless it received a blood transfusion. He decided to take a chance and use the trusting and gentle Ramone as a donor.

For two minutes, while Dr. Otto stuck a needle in Ramone's jugular vein and slowly withdrew a fourth of a cup of blood, the good-natured cat sat quietly and patiently. Veterinarians don't have to worry about matching blood types in felines, and Dr. Otto immediately completed the transfusion. The other cat was so weak from the disease that it didn't protest the insertion of the needle with the life-giving fluid into its body.

Dr. Otto said that the great majority of cats would quite likely go berserk if he were to stick a needle in them, especially in their jugular veins. But he has never even had to sedate the uniquely benevolent Ramone, who seems somehow to sense that he has been responsible for saving the lives of many of his fellow felines.

*I*n the fall of 1968, a tomcat named Thug received an award for heroism from the Los Angeles Society for the Prevention of Cruelty to Animals after rescuing a dog. Normally an exceptionally quiet cat, Thug spotted a Labrador retriever being swept under a pier at a marina. The tom began a series of alarming yowls that drew attention to the dog's peril, and Missy the Labrador was saved.

J. M. Escobedo of La Jolla, California, told us about the strange relationship that his large gray tomcat, Big

Boy, had with his old rat terrier, Taco.

Taco was already ten years old when the family acquired Big Boy. At first the two fought like the proverbial cat and dog, but as they grew older, a kind of grudging affection developed between them. In 1996, when Taco went blind because of cataracts, Big Boy completely stopped teasing and baiting the dog and seemed to become his protector.

"One afternoon, I was sitting reading the newspaper on our veranda, only vaguely aware that Taco was wandering out of the yard," J. M. Escobedo said. "Suddenly Big Boy dashed from beside my lawn chair where he had been nibbling contentedly on a midday snack and let out a loud cry. He saw what I had not. Taco, no longer able to see, was heading for the steep incline that dropped off to the beach below. If he toppled over the edge, he was certain to have a very bad fall.

"Big Boy caught up with Taco just in time. The cat placed its body directly in front of the old dog, and Taco stood still, his short tail wagging in frustration over his inability to see. Taco whined, and Big Boy meowed in answer.

"It was weird and wonderful at the same time," Escobedo concluded his account." Slowly, moving from first one side and then the other, Big Boy carefully nudged Taco along and brought him safely home."

*I*n 1991, when we moved to the newly developed desert housing units outside Cave Creek, Arizona, we did not fully comprehend that in this particular region the West was still wild in many ways.

Within the first week of our move, we encountered a six-foot-long rattlesnake on the driveway, a scorpion in the dishwasher, four or five lizards in the bathroom, two hairy tarantulas on the patio, a pack of six coyotes snarling at our Black Labrador in the backyard, and a herd of thirteen javelinas, wild pigs, in the front yard, which loved to look in our daughter Melissa's window.

We found one feature of our new surroundings, the mournful moonlight dirge of the coyotes, strangely satisfying. But then we found out how wily those coyotes could be. We heard from neighbors who had their cocker spaniel eaten by a pack of coyotes.

The coyotes had worked out a clever and vicious method of both combating and devouring the canine competition in the desert. The pack would cleverly send a female in heat up to homes in which they could detect the scent of a dog dwelling within. When the love-struck hound went out in pursuit of the lady coyote, the male coyotes would jump him, kill him, and share his body with the hungry pack.

We kept a protective watch over Moses, our Black Lab, but we learned that dogs are not the only victims of the coyotes. It seems that a coyote loves nothing better for a midnight snack than a cat that didn't make it home before dark.

From another neighbor in Cave Creek, we heard the legend of the mighty Toro, a cat that fought back when a couple of coyotes tried to gulp down a fellow feline as a fast food treat.

In the mid-1980s, Gus and Margaret Holt moved to Cave Creek with their two kids, Tammy, nine, and Jason, seven, and their big black tabby, Toro.

At first the Holts were continually concerned about the kids being bitten by any one of the assorted wildlife of the desert and about Toro, who they feared might try to tease in a catfully playful manner some poisonous creature such as a rattlesnake or a scorpion. While Tammy and Jason lived in dread of whatever might be the creepy-crawly of the day, Toro suddenly found himself the possible diet of choice for a large variety of predators.

As the Holts settled into their new home in Cave Creek, the practical-minded Gus began to see the various hazards of the area as a mixed blessing. Although there was a definite threat to Toro and possible dangers to the kids, Tammy and Jason suddenly became much more responsible in terms of seeing that their pet was home safely. In their old neighborhood in Scottsdale, the kids had been pretty lax about whether or not Toro was safe at home after dark.

Gus had always known that Toro was a "macho" cat, reluctant to back down before the most aggressive dogs in the neighborhood, but somehow the big tabby had to understand that coyotes were much hungrier than the average urban mutt. As his son Jason had phrased it, if Toro had a hero, it was probably Heathcliff, the pugilistic tiger cat in the Sunday comics.

It was Tammy who actually witnessed Toro's act of heroism.

She and Paula, who was also nine and who lived next door, were playing dolly dress-up with their cats in the Holts' backyard. Paula had pulled a pink dolly sundress on Ramona, her Persian, and Tammy was trying to put a yellow apron around Toro's big belly. The two girls didn't even see the two coyotes until they were right in front of their noses, growling at them and showing their teeth.

Although their parents had warned Paula and Tammy about the boldness of coyotes in invading garages and yards, the adults had never anticipated that the scavengers of the desert would have the audacity to approach human beings, even little ones such as their daughters.

Tammy said later that Ramona, Paula's Persian, had started yowling and hissing at the uninvited guests, but that Toro had just studied the intruders closely, as if looking for an opening.

Ramona's defiance either provoked one of the coyotes or stimulated his hunger, for it lunged at the cat and snatched her up in his jaws in one practiced swoop.

That was when Toro went into action.

Tammy reported that the brave tabby made a really big jump and landed on the head of the coyote that had grabbed Ramona.

The coyote dropped Ramona and shook Toro off. Then Toro went right for its face, making a terrible howling while he scratched and scratched at the coyote's nose.

The coyote yelped so loud that both of them turned and ran from the yard. Toro just stood there, keeping an eye on their backsides as they disappeared in the desert.

Somehow, according to Tammy, he almost seemed

disappointed that the battle had ended so quickly.

Ramona required a number of stitches at the veterinarian's, but she was not much the worse for wear by dinnertime that evening. Paula's parents bought Toro a can of expensive cat food as a gift for saving their daughter's pet.

The big black tabby had joined the ranks of the bold and courageous and become one of the first feline heroes of the Southwest.

*A*lthough our nation's first president, George Washington, had a number of cats with such names as Truelove, Sweetlips, Mopsy, and Madam Moose, none of them made it to the White House for the simple reason that the presidential residence was not ready for occupancy until 1800, three years after Washington completed his last term in office.

Abraham Lincoln's son Willy owned a cat that soon gave birth to a whole litter of kittens that the president took great delight in helping to name. According to available historical records, the Lincoln litter (circa 1862) were the first cats in the White House.

When Rutherford B. Hayes was in the presidential residence from 1877 to 1881, his gift of a Siamese cat from the U.S. consul in Thailand was the first of that breed in the country.

President Theodore Roosevelt, who is usually associated with big game animals, nonetheless had a six-toed cat named Slippers, who enjoyed a run of the White House from 1901 to 1909. More in keeping with his rugged reputation, Teddy also at various times gave presidential refuge to a bear, a lizard, and a snake.

Woodrow Wilson, the tireless champion of world unity who won the Nobel Peace Prize in 1918, tried his best to break his cat Puffins of its penchant for chasing birds. According to the story, one day what appeared to be an entire flock of birds swooped down on the feline predator and gave him such a good pecking that Puffins never bothered another bird from that time on.

Calvin Coolidge proved during his time in office from 1923 to 1929 that he was not such a dour fellow after all, when people saw the affection that he displayed for his cat Tiger. Once when the cat wandered off, Coolidge took advantage of his office to prevail on all the Washington, D.C., radio stations to put out all-points bulletins to locate Tiger. Fortunately for cat lover Cal, Tiger was found and returned to the White House.

In addition to Tiger, Coolidge's favorite, the family

kept a number of other cats, an aviary of birds, numerous dogs, and a raccoon named Rebecca that Mrs. Coolidge liked to walk on a leash.

President William Jefferson Clinton, first lady Hillary, and daughter Chelsea brought their cat Socks to the White House. Clinton often expressed his admiration for President John F. Kennedy, and he also shared JFK's allergy to cats. While little Caroline Kennedy, like Chelsea Clinton, did own a cat, her Tom Kitten was sent off to live with the first lady's press secretary. Caroline continued her long-distance relationship with Tom Kitten for years and came to visit his grave and pay her respects after he died.

When Tim and Kristi Henniger moved to a new neighborhood in a suburb of New York in 1998, they were astonished by the laxity of dog owners who failed to keep their pets in check.

"Complaining brought apologies and promises to keep Fido on a leash," Kristi said, "but it just never happened."

"The doggie-do in the yard and on the sidewalk was one thing," Tim added, "the urine stains on the tires and fence posts were another. I planted two rosebushes, then lost them to the mutts within the first four days after we moved into the neighborhood."

The Hennigers had lived there only a month when their daughter Tina asked them to take care of her cat, Carlotta, a cross-bred Maine coon cat and Persian. At first the couple hesitated to assume responsibility for Tina's big brown cat. After all, the neighborhood was crawling with dogs. What if one of them injured or killed Tina's beloved companion? Would their daughter blame them for not keeping a closer watch on Carlotta?

"The other reason for our hesitation, quite frankly," Tim said, "was that I was allergic to cats. And besides that, I had never really cared for the things."

The Hennigers finally consented to take Carlotta for the next two weeks, and Tina dropped the cat off on her way to La Guardia Airport.

"You'll pardon me if I leave the care and keeping of this big feline primarily to you, dear," Tim said to Kristi, reaching for his handkerchief and stifling an exaggerated sneeze. "My allergy, you know."

The next afternoon, Kristi was working on her word processor and listening to Beethoven on their CD player when she glanced out the kitchen window and spotted a large German shepherd boldly approaching Carlotta, who was sunning herself in the backyard.

Kristi grimaced and thought to herself that poor Carlotta was about to be converted into mincemeat. Tim was at work, and she was very cautious about running afoul of an angry German shepherd. Kristi was already

🐾 🐾 🐾 *Cat Miracles*

trying to decide what she would tell Tina about Carlotta's fate when her daughter returned from her vacation.

Carlotta barely opened her sleepy eyes, reached up, and expertly raked the shepherd's nose with the extended claws on her left paw. The dog ran yelping from the yard.

Within a few days, it became abundantly clear that Carlotta was the equal of any dog in the neighborhood. Cocker spaniels, poodles, and beagles were easy knockouts, and she barely ruffled her fur taking out the mean rottweiler from around the corner and the German shepherd, which returned for a rematch.

"It's amazing," Tim laughed when he returned from work on the fourth day of Carlotta's visit. "The dogs in the neighborhood are starting to avoid coming near our yard. Old Slugger Carlotta is scaring them all away. I think I'll try planting rosebushes again."

On Saturday afternoon, the seventh day of the reign of Carlotta, Warrior Queen, Tim was able to observe their tough cat in action as he tamped the soil around a freshly planted rosebush.

"A big rottweiler approached our yard," he said. "I had seen the brute in the neighborhood, and I knew I wouldn't want to tangle with him. He walked up to the edge of our grass, and he had this low, throaty growl building in his throat, saliva was dripping from his mouth.

"Without a moment's hesitation, Carlotta jumped up from where she had been napping beside me and came hissing and spitting at the big rottweiler. The monster backed up, but then decided to stand his ground. That was when Carlotta moved deftly between his front legs and sank her fangs up to the hilt in one of his back legs. The brute yelled all the way back to his own yard around the corner."

That night, Tim gave Carlotta an extra helping of her favorite mixture of cheese and tuna, then conducted a formal weighing-in on the bathroom scale.

"It might not be official," he told Kristi, "but our champ tips the scales at nearly twenty-eight pounds. No wonder she has no fear. She knows she's the biggest, meanest tiger in the valley."

On the tenth day of Carlotta's stay with the Hennigers, Kristi noticed for the first time since they had moved to the neighborhood that the dog owners were walking their pets on leashes. She also observed that most of the dogs whimpered as they were tugged past the edge of the Hennigers' front lawn.

"Carlotta has all the neighborhood dogs running scared," Kristi reported to Tim that evening when he came home from work. "Now nearly everyone has their dog on a leash, and Mrs. Bernstein from across the street called to see if Carlotta was inside the house before she would walk her dog."

Tim bent down to give Carlotta an affectionate scratch behind the ear. For some reason, his allergy to cats no longer seemed to bother him.

"Carlotta is like Wyatt Earp cleaning up Dodge City," he said. "She has the fastest claws in the West."

The next morning when he was leaving for work, Tim was amused to see that some neighborhood wit had placed a crude "Beware of the Cat" sign next to their front walk.

When the two weeks of Tina's vacation were up and she stopped by to pick up her "sweet little kitty," Tim told his daughter she could have Carlotta back only if she promised to bring her over for a visit every few weeks. He wanted to plant some more rosebushes, and he didn't want the dogs in the neighborhood to forget the vengeful presence of Carlotta, Warrior Queen.

*A*s the official rat catcher on board the British vessel HMS *Gorleston*, Rastus made the forecastle of the ship his special area to take a little relaxation time from his duties in the hold. On this particular day when Rastus was lounging in the sun, a very large seagull landed near him and appeared ready to dispute the true ownership of the forecastle.

Rastus rushed the winged intruder who had dropped in uninvited from the sky, growling and hissing in his most fierce demeanor. The seagull stood its ground, unimpressed by Rastus's hostile advances, fully ready to show the cat the might of his large beak and the power of his beating wings.

Realizing that the large bird was not to be easily intimidated, Rastus sunk his fangs into one of the seagull's legs. Let's see how the squawking bag of feathers handled that move, Rastus may have thought, already declaring himself the victor.

Rastus may have been correct in assuming that a solid bite to the leg would drive the intruder away from his special lounging area, but he was not at all prepared when the big seagull suddenly shrieked and took off with Rastus and his mighty jaws still attached to its leg. As astonished crew members of the *Gorleston* watched in amazement, they beheld their ship's cat flying through the sky above them solidly clamped onto a seagull.

Rastus soon decided that he would leave airborne flight to creatures with wings, and he unclamped his jaws and plummeted to the sea below. Fortunately for Rastus, his encounter with the seagull took place while the ship was docked in Liverpool harbor and not while it was moving rapidly through the North Atlantic.

As crew members cheered him on, Rastus swam to the shore, crawled up on the dock, and reboarded the ship. As he walked, soaking wet, back on board, crew members noted that Rastus acted as if the entire incident had been played out just as he had planned it.

The story of Rastus and several other cats that accompanied the British fleets in peacetime and in war was recorded by J. D. Carpentieri in the August 4, 2001, issue of *The Guardian*. Although some cultures consider it bad luck to bring a cat aboard a sailing vessel, it would appear that most British seamen take the opposing view. Because they believe a cat on board to be a harbinger of good fortune, the crew of the HMS *Cossack* made a special effort to rescue a cat from the famous German battleship *Bismarck* when British forces sank it in the spring of 1941. The crew named him Oscar and treated him as a very special prisoner of war, but the cat didn't bring them good luck.

Within six months of their participating in the sinking of the *Bismarck*, the *Cossack* was sunk by a German submarine. Oscar, however, was found calmly cleaning himself on a piece of wood floating amid the ship's wreckage, and he was picked up by crew members from the HMS *Ark Royal*. This ship fared no better in receiving any blessings of good luck from the unsinkable Oscar. It was sunk only three weeks after rescuing the cat from the sea.

While some seamen may have suspected Oscar of being an active saboteur for the Third Reich, the cat was placed in an animal shelter in Belfast, rather than receiving a court-martial and the firing squad for being a spy.

The Rabies Control Order of 1974 ended the practice of bringing on board a ship's cat, for pets could no longer enter the United Kingdom without undergoing a period of quarantine. The last of the legendary feline seafarers was Fred Wunpound, who in his nine years of service aboard a British ship logged over 250,000 miles at sea, an average of one full trip around the world per year.

*D*uring their summer vacation in 1993, John and Cassandra Kraven headed for a cottage in the Adirondack Mountains in upstate New York with their two-year-old daughter, Jane, and their cat, Sidney. Actually, they had not wanted to bring the cat along, fearing that it would be too much bother looking after it. They were concerned that their city-bred cat would be confused and wander off in the woods, where a predator might make a quick meal of him. However, a friend who was supposed to kitty-sit was called out of town at the last minute, so

Sidney, his litter box, and several cans of his favorite cat food were packed in the car.

The first few days at the cottage had been truly relaxing, but their tranquility was nearly irreparably shattered when a black bear suddenly stormed into their yard. Before John or Cassandra could respond to the situation from within the freeze frame of shock that had enveloped them, the huge animal had grabbed their daughter and was shaking her in his snout as if she were a rag doll.

At that moment of ultimate horror, Sidney leaped onto the bear's head, fastened his back claws into its flesh, and scratched at the brute's eyes until it dropped the baby in order to better direct its wrath at the attack cat.

Sidney, beholding a momentary victory with Jane now released from the behemoth's jaws, jumped to the ground and deftly avoided the clumsy giant's swiping paws. Then, with the bear in hot pursuit, Sidney dashed into the forest.

John and Cassandra ran to their child, who, though crying in terror, seemed unharmed. Miraculously, the bear's teeth had snatched the girl's playsuit and had not punctured her flesh. If it hadn't been for Sidney's dramatic intervention, however, they were horrified to think of what might have happened.

After about two hours, Sidney returned to the cabin, completely unharmed. The Kravens theorize that their courageous kitty led the bear on a merry chase deep into the forest, far away from their cabin, in order to ensure the return of peace and quiet for the remainder of their vacation.

John and Cassandra know that they owe their daughter's life to Sidney, the cat that they wanted to leave behind. They said that they will never visit their cottage again unless they are in the company of their fierce feline protector.

*M*any cat lovers were annoyed with what they considered the stereotypical typecasting of felines as sinister, secretive, and subversive in the motion picture *Cats & Dogs* (2001). The film plays upon what appears to be the age-old rivalry between domesticated dogs and cats and portrays this tension as an eternal struggle between two warring armies for domination over each other. Mr. Tinkles, the evil Chief Cat (voiced by Sean Hayes), plans to divert the research of Professor Brody (Jeff Goldblum), who is working to achieve a vaccine to rid humans of dog-hair allergies, and alter the formula to render all humans overpoweringly allergic to all dogs. In Mr. Tinkles's evil plot, all dogs would therefore be exterminated by humans, and cats would rule the world. It falls upon a beagle pup (Tobey Maguire) and other members of the canine corps to foil the sinister feline conspiracy to achieve world dominance. While live animals were utilized for many scenes during the filming of *Cats & Dogs*, the close-ups and the accurate movements of the lips for speech were achieved by puppets, animatronics, and sophisticated special effects.

Movie fans can name numerous canine cinema stars such as Strongheart, Rin Tin Tin, Lassie, Bullet, and Benji, to list only a few. There are few cats in the Hollywood Hall of Fame because of the difficulty of training a cat to do tricks on demand.

Cats have fared little better in the comic strips than in the movies. While Felix was a popular heroic figure of a feline both in the comic pages and in animated cartoons for decades, Krazy Kat, pretty much identified by his name, may have been a kind of forerunner of Sylvester, who is always trying to sneak up on Tweety Bird.

George Gately received his inspiration for the pugilistic cartoon cat Heathcliff from his brother's tough, independent tabby, Sandy. "All the dogs in the neighborhood were afraid of him," Gately recalled once in an interview. "Sandy would fight anything on four legs—a lot like Heathcliff."

Garfield's creator, Jim Davis, had twenty-five cats

when he was a kid, but doesn't have one now because his wife is allergic to them. The cartoon cat is based on a variety of his friends' cats, Davis said. "I grew up on a farm and cats were all over the place. The fact that I don't have one now means I use other people's cat stories and Garfield is more of a general kind of cat."

However, getting back to the movies, Rhubarb was a pretty big cat star in *Rhubarb* (1951), the story of a cat who inherits a baseball team, but the incredible Syn Cat was declared by animal trainer Al Koehler to be "the smartest, most sociable, most emotionally stable cat in the world." Proclaiming Syn Cat, the affable Siamese, to be a feline that only comes around once in a lifetime, Koehler saw Syn Cat go on to star in such memorable Disney films as *Incredible Journey* (1963) and *That Darn Cat* (1965) with Hayley Mills and Dean Jones.

And before there was the finicky Morris hustling cat food in all those television commercials, there was the elegant and stylish Nicodemus, who lived a true ugly duckling story. The unwanted runt of a pedigreed litter, Nicky grew to be the glamorous, snowy white Persian who became a famous model for Revlon cosmetics. Nicodemus, in fact, went on to become a cottage industry; the sophisticated cat even guested on the *Today Show, Captain Kangaroo, Play Your Hunch,* and many other television shows.

The success story of Princess Kitty offers another alley-to-riches story. In 1986, she was a six-month-old stray kitten with a BB shot in her body who wandered into the yard of Karen Payne of Miami, Florida. Eight years later, Princess Kitty was being hailed as the smartest cat in the world, recognized for being able to perform seventy-five tricks, including counting, jumping through hoops, slam-dunking a basketball, and even playing the piano.

Karen Payne said that after the veterinarian gave the abused kitten the necessary injections to restore health, he told her he could tell that the cat was very intelligent. She took his advice and began to teach Princess Kitty some tricks. Within hardly any time at all, the kitten was fetching, shaking hands, and jumping through hoops.

After Princess Kitty had mastered dozens of stunts and maneuvers, Karen bought a miniature piano with keys spaced far enough apart for a cat's paw to play. With Karen pointing to the keys with a conductor's baton, Princess Kitty can play "Three Blind Mice" and the opening bars to Beethoven's Fifth Symphony.

Soon Princess Kitty was performing for children in schools and hospitals, and in 1988, she appeared with actor Stacy Keach in the television miniseries, *Hemingway*.

In 1994, Alan Beck, a Purdue University professor of animal ecology, observed Princess Kitty performing and commented that she was like a little human in a cat's body. A cat such as Princess Kitty, who can learn tricks easily and retain them, is extremely rare.

While there are approximately 63 million cats in the United States that are regarded as family members and properly fed and cared for, another approximately 60 million run wild in both urban and rural areas and become the victims of dogs, coyotes, bobcats, hawks, and human beings. Each year, game wardens justify trapping or shooting thousands of these feral (wild) cats by arguing that they prey upon domestic stock and game animals. Fish and Wildlife Services state that they eliminate feral cats only when they suspect the creatures might have rabies, but FWS brochures recommend the destruction of all vagrant cats.

While there may be justification in destroying those feral cats that present a danger to humans or other animals, cruel and inhumane treatment to such creatures can never be justified. Stubbs, a black-and-white kitten of mixed breed, was living wild in Los Angeles when some thoughtless individuals took it upon themselves to kill the wanderer by dousing it with lighter fuel and setting it on fire.

The pain had to have been excruciating, but somehow Stubbs managed to escape his would-be assassins by running into an alley, and dousing most of the searing flames by rolling on the pavement. The cruel individuals, satisfied that the kitten would eventually die from the burns, walked away from the injury that they had afflicted on their tiny victim, perhaps even amused by the sounds the kitten made as it cried out in pain.

Stubbs lived, limping from place to place on legs and toes that had been terribly burned by the lighter fluid. After several days had passed, his left front leg and the right rear leg had become infected. Within a few more days, they had become dangerously gangrenous. Somehow the kitten understood that the poison in those two legs would kill him unless he did something about it.

Displaying incredible courage and resolve, Stubbs set about the painful process of chewing through his left front leg near the knee and the right rear leg at the knee.

It was shortly after the brave kitten had chewed off part of two of his legs and a toe on his right front leg that an animal lover happened to come upon Stubbs. Carefully picking up the burned and mutilated cat, the humane individual took him to the Woodland Hills Veterinary Clinic.

Dr. George Grant, the animal surgeon who treated Stubbs, said that the cat had saved his own life by self-amputating the gangrenous portions of his legs. Joanna Patrice, spokesperson for Cat/Canine Assistance, Referral and Education (C.A.R.E.) of Sherman Oaks, continued to look after Stubbs during his recovery and was soon pleased to state that the tough little guy was learning to trust humans and allowing himself to be petted. Extra good news was that the veterinary surgeons were about to give Stubbs artificial limbs to replace those that he knew intuitively had to be re-moved in order to save his life.

*L*iterary scholars and Lewis Carroll enthusiasts alike have long pondered the origin of the author's inspiration for Wonderland's Cheshire cat. From what source did he derive the bizarre feline that grinned at Alice until only its smile remained?

In July 1992, literary detective Joel Birenbaum discovered the original Cheshire cat lurking in the shadows at St. Peter's Church in Croft, Durham, England, and the longstanding mystery of where Lewis Carroll got the idea of a cat that disappeared leaving only a smile behind was solved at last.

One of thirty-five members of the Lewis Carroll Society visiting the church as part of a pilgrimage to various locations frequented by the revered author, Birenbaum observed a most curious phenomenon. As he knelt at the altar of St. Peter's, he noticed the image of a cat that had been crudely carved in relief on a stone wall panel.

As he knelt lower, he was astonished to perceive that the lighting in the tenth-century chapel created the illusion that the cat disappeared—except for its smile.

When people look at the image of the cat from the front, it looks just like the image of an ordinary cat. But, Birenbaum illustrated for his fellow Carroll enthusiasts, when people go down on their knees and look up, they can see only the grin and not the full representation of a cat.

Carroll, whose birth name was Charles Lutwidge Dodgson (1832–1898), moved with his family from Cheshire to Croft when he was a boy of eleven. It seems likely, as Birenbaum has suggested, that the lad became intrigued by the illusion of the disappearing cat as he knelt in the chapel of St. Peter's Church. The peculiar effect apparently so delighted him that he used the lingering feline grin as an attribute of the Cheshire cat many years later in his famous *Alice's Adventures in Wonderland* (1865).

*I*n November 2001, Victor
Marchetti, a former officer
with the Central Intel-
ligence Agency (CIA), released details from a document
recently declassified from the Science and Technology
Directorate, which described one of the most astonishing
surveillance devices created during the Cold War: a cat
with a transmitter placed in its body and an antenna
inserted in its tail.

In 1966, the CIA devised an elaborate plan that called
for a cat with built-in bugging devices to stroll noncha-
lantly by Soviet agents as they sat on benches in a

London park and pick up details of their conversations. Hopefully, the Russians, thinking they were safe from the prying ears of Western intelligence agents, would freely discuss Kremlin secrets, unmindful that the innocent cat strolling by them was picking up their every word.

The CIA technicians anesthetized the cat, whom they called Acoustic Kitty, slit him open, placed batteries within his body, and wired him up. The antenna for the receiving set within his body was placed in Acoustic Kitty's tail.

After a great deal of trial-and-error testing, the CIA felt that Acoustic Kitty was ready to spy for the greater good of the West. Two Russian agents were known to frequent a particular park during their lunch break, and the CIA assumed that they would be engaged in shop-talk as they ate lunch on their favorite bench.

The CIA van, complete with state-of-the-art listening devices, parked discreetly some distance away from the Soviet agents and released Acoustic Kitty, loaded with bugging devices, to do his duty.

Marchetti concluded his remarkable account by revealing that Acoustic Kitty was barely out of the van and headed toward the park bench when a taxi ran over him.

The CIA agents sat stunned in the van, surrounded by all their sophisticated surveillance equipment, and saw weeks of experimentation, as well as Acoustic Kitty, wasted in one awful moment.

On that spring day in 2000 when Brad opened the front door of our residence in Iowa to see what was making a yowling noise outside the house, he found a medium-sized peach-colored cat on the front steps. We had recently lost our beloved Moses, a sturdy Black Labrador, and we hadn't made any attempt to acquire a new animal friend, so Brad was not particularly enthusiastic to find a cat asking him for a handout. He knew that he was a sucker for a fuzzy-faced feline, especially one that could so quickly transform a yowl for attention into a soft, melodic meowing.

In a straightforward, almost grumpy, tone of voice, Brad spoke to the cat: "All right, I'll give you something to eat. But everyone works for a living around here, my friend. If you think you're going to stay around the place, you'll have to pull your own weight."

Brad gave the cat a bit of tuna that was left over from lunch and went back to writing, pretty much dismissing the feline from his mind, assuming that after its free lunch, it would be on its way. He had noticed that the cat was a neutered male, so it was obvious that it had once belonged to someone. Perhaps it would return to its home after it had enjoyed the snack that Brad had provided.

Brad was astonished the next morning when he opened the front door to go to the mailbox to discover that the cat was waiting for him on the front steps. Proudly, the fellow meowed and indicated the heads of a rat, a field mouse, and a mole, neatly arranged as trophies before him.

Brad called me to witness the exhibit of hunting prowess displayed for our approval by the accomplished hunter. We agreed that he had proven his worth as a stalker of rodents around our country home. We told him that he was welcome to remain to patrol the yard if he wished, and we named him, then and there, Bart the Bold.

Once the negotiations with the cat were completed, Brad continued on his way. Bart followed him to the tin mailbox atop the wooden post, keeping a pace or two of respectful distance between them. The daily trek down the lane to check the mail would become a ritual that would begin the day for Bart and Brad for the two years that the cat remained with our family.

One afternoon a few weeks after Bart had arrived to join our homestead, Stan and Ruth, the farmers who once owned the pasture next to the river on which our house presently stands, stopped by to visit. Ruth laughed in warm recognition when she saw Bart, and she picked him up to pet him.

"You call him Bart," she told us. "I call him Peaches." Or, Ruth went on to explain, he was a descendent of Peaches, at any rate. Perhaps a great-great-great-grandson—or maybe even a few more "greats," besides.

Peaches had been a female cat of Stan and Ruth's that had gained a local reputation for intelligence. In fact, Ruth told us, Peaches had been so bright that when the college drama department needed a cat for a play, they asked to borrow Peaches. With a proud smile recalling the fond memory, Ruth stated that Peaches never missed any of her cues during the run of the play.

We were pleased to learn what distinguished ancestral blood flowed in Bart's veins, but we also understood that he must have at one time been one of the feline barn dwellers at Stan and Ruth's farm. Perhaps he would decide to follow them back home, and we would lose our little peach-colored rodent catcher. If that were to happen, we knew that we would miss our friend.

Bart jumped down from Ruth's arms, and he walked a few paces with Stan and Ruth down the lane. Then he turned and walked back to stand beside us. It was as if Bart was seeing his old benefactors down the lane, indicating that it was good to see them again, but he was happy in his new home.

Throughout that summer, Bart patrolled the riverbanks, checked the sheds, and roamed through our acreage, keeping the rodent population in check.

Once he made a dash for a squirrel that was carrying an ear of corn from a nearby field, but the quick-moving, bushy-tailed fellow dropped his booty and sprang for the trunk of a large oak tree. Bart was a remarkable tree-climber, but he was no match for a creature that made the high branches its home.

Amidst the sounds of the squirrel's scolding chatter, Bart jumped back down to the grass and walked away from the tree with the studied indifference that cats mastered centuries ago. He stopped just once to look up at the noisy squirrel dismissively, as if to say, "If you stick to the trees, I'll leave you alone. But if I ever catch you messing with things around the house. . . ."

That fall, our friends Jon and Laura made the decision to move out of town, and they asked us if we would like two small kittens to keep Bart company. We said we would give the two kittens a home, although we were uncertain how Bart would take to new arrivals on his turf. Our previous experience with male cats led us to believe that male cats aren't especially nurturing and don't always make the best caregivers to kittens.

But once again, Bart proved to be the exception. He immediately accepted the two small black fuzz balls with white boots and collars, and he even indicated that he had no problems with their eating out of his special bowl. In fact, from their very first meal on, Bart always let them begin to eat while he kept a wary eye for interlopers such as the possums and raccoons that sometimes invited themselves to dinner.

We named the kittens Freya and Lena. Then, after a few weeks, the two little girls underwent name changes to Fred and Leonard. (Sometimes it is difficult to determine the sex of very small kittens.) The two guys quickly fell into the workaday schedule, even joining Brad and Bart in walking down the lane when it was time to check the mailbox.

As the nights grew colder and winter was setting in, I began to winterize the large doghouse that a previous owner had left on the property. Perhaps "small house" is a better description, for any adult under six-foot-six can stand upright in it with no difficulty.

I am allergic to cat dander, so it was not possible to bring the three cats into the main house, but I made the accommodations in the doghouse so comfortable that there were no complaints from the three boys. There were snug beds with lots of old blankets to keep them warm, and for really cold nights, there was even a heat lamp to raise the temperature and keep the house nice and cozy.

One night as Brad entered the little house to feed the cats, he was startled to see that they had a very unwelcome uninvited guest and that Bart was locked in a staring contest with it.

At first glance, Brad thought that a bobcat had invited itself for dinner — which would likely include Bart, Fred, and Leonard — and he froze in his tracks. Many years before, Brad and his college roommate had encountered a bobcat when they were rock climbing, and he had never forgotten coming face-to-face with that large, angrily hissing wild cat. Fortunately, Brad's friend had had a pistol and had frightened the bobcat away. Brad wished that he had a pistol in hand at that moment.

Upon closer scrutiny, Brad decided that it was not a bobcat, but a very large feral cat that had invaded the cats' castle. Feral cats are domestic cats that have gone wild, and their offspring, which are born in the wild, have had no exposure to humans during the critical socialization period when they are from three to eight weeks old.

Knowing that he now had Brad's physical support, Bart emitted the loudest, eeriest, screeching yowl that one could ever imagine might come from such a small cat. In the next moment, he rushed the feral cat and chased it from the house and into the darkness.

Brad heard the unmistakable sounds of a cat fight taking place somewhere amidst the trees, and he was concerned for Bart, for he was less than half the size of the feral cat.

Brad called for me to bring a flashlight so we might find and rescue Bart, but the flashlight beam soon picked up the form of a triumphant Bart returning to the little house, seemingly none the worse for wear.

However, on the next night when Brad went to feed the cats, the invader had once again entered the domain of Bart and the kittens and was crouching in the shadows, held back from the crumbs in the bowls by Bart's unwavering glare. It had snowed heavily the night before and there were drifts four feet high all around the little house. Feeling compassion for the hungry cat, Brad distributed fresh food in the bowls and placed some extra in an old frying pan. Then he told Bart to let the cat eat.

Bart was hardly a selfish animal. He had proven that fact countless times when he generously allowed Fred and Leonard to eat pretty much their fill before he would even begin his meal. But he remained unfriendly and alert toward the feral cat. He positioned himself between the kittens and the uninvited guest and watched the intruder cautiously as it ate from the pan.

The next evening turned cold, and I went to the cats

to feed them and to be certain the heat lamp was turned on to keep them warm. Spotting the feral cat, I made an overture of hospitality to it.

It was fortunate that I had worn an extra-heavy coat that night, for the wild cat leaped on my arm and sunk its teeth into the coat. Its jaws began to grind against the material, deliberately seeking to inflict serious harm on me.

That was when Bart threw himself on the vicious stranger, seized the feral cat by its throat, and began trying to drag it off of me. Although I said later that Bart looked like a little cowboy trying to ride a Brahma bull, he would not be shaken away by the much larger cat. He fought with all his heart and might to protect me, and he finally pulled the feral cat from my arm and chased it into the night.

I called Bart my little hero and sprinkled some extra food into his bowl. Bart, still panting from the violent encounter, simply stood back, as always, and watched over Fred and Leonard while they ate. But he had proven once again that he was continuing to fulfill the contract he had agreed upon when he arrived on our front steps, and was definitely pulling his own weight.

Bart remained our loyal "watchcat" and a protector to Fred and Leonard until they reached their maturity. Then, in that secret way that only very wise cats can understand and well-intentioned humans can only guess,

Bart heeded a silent call to leave this earthly life and enter the Great Mystery.

Knowing Bart, however, we wouldn't be surprised if his spirit returned one day in the body of another cat, meowing at our front door and negotiating for a position as rodent catcher.

A millionaire we'll call "Jack" owns a string of sandwich shops in Toronto. When he filled out an application for a credit card in the mid-1990s, he really didn't foresee any problem in obtaining a piece of the precious plastic. When he was rejected, he was astonished.

Here he was, thirty-four years old, a successful entrepreneur with an estimated net worth of one million dollars. How hard could it be to get a credit card?

Someone—Jack swears he doesn't know who—overheard his complaint and thought it would be a good joke to fill out an application for a credit card with another bank in the name of Munchie, Jack's pet cat.

Incredibly, almost by return mail, Munchie, a three-year-old black cat, got her credit card. So Jack went out and bought some kitty litter on her new card, a plastic passport to comfort.

Bank officials insist, however, that they did not extend credit to a cat, but to Jack, who gave the cat authority to use his account.

Boris didn't need a credit card to order 450 cans of his favorite chicken cat food on an Internet shopping Web site in November 2001.

According to his owner, Betty Richards of Cambridge-shire, England, Boris happened to walk across the computer keyboard after she had just ordered six cans of cat food. While she wasn't looking, Boris somehow managed to strike the keys that transformed the order from 6 tins to 450.

Betty didn't realize what the cat had done until she saw that the order totaled $684.00. She said that it just had to be a fluke that Boris's paws just happened to hit the right keys—but perhaps her cat may be more clever than she can imagine.

*B*abi and Kuukie took three trips from New York to Israel in the year 2002 and became the first cats to earn frequent flyer miles from a commercial airline and to be awarded a free flight.

Their owner, Elayne Rifkin from White Plains, said that she and Babi and Kuukie average four trips a year, because the cats were born there and they love to visit their homeland. Israel's El Al airline recently instituted a frequent flyer program for pets.

Then there's Ozzy, who became a frequent flyer by default when he escaped from his plastic basket and clocked 63,000 air miles in ten days.

Ozzy's owners, Jonathan and Katie, were flying with him back to the United Kingdom from Qatar when he somehow managed to squeeze out of his basket and was absent without leave when the British Airlines staff opened the hold at London's Heathrow airport. His owners assumed that Ozzy must have disappeared during a stop in Bahrain, and they reconciled themselves to the sad fact that they would never see him again.

Ten days later, however, Ozzy was sighted in the hold of the British airliner. He had flown undetected in the hold and had made the 6,300-mile round-trip flight every day. After the frequent-flying feline had been returned to his grateful owners, a British Airline spokesperson said that if Ozzy had been registered for BA Miles, he would have earned enough points to travel free to Rio de Janeiro and back, plus a couple of trips to Europe.

*J*essie Neal told us that she had just let Lady, her dachshund, out into the yard of her home in a suburb of Atlanta on a sunny afternoon in September 1999, when the tiny dog was attacked by a massive pit bull that had invaded the Neals' property.

"I had never seen that monster before," Jessie told us. "It just seemed to come from out of nowhere and grab poor Lady in its huge jaws. It acted like it wanted to make a meal out of her."

Lady was being severely mauled when Chang, the Neals' Siamese cat, leaped from the roof of a nearby porch to land squarely on the head of the savage pit bull.

"My husband had named Chang after some kung fu master in one of those martial arts films that he loves to watch," Jessie said. "And that cat certainly earned its name that day. It just tore into that monster dog like Bruce Lee or Jackie Chan mopping the floor with the villain."

Nine-pound Chang drove off the vicious dog and saved the dachshund's life.

"If Chang hadn't been there, Lady would have been hamburger," Jessie Neal said. "I called the people at the dog pound to come looking for the pit bull. It didn't belong to anyone in our neighborhood, I knew that for certain.

"One of the strangest parts of this whole incident," Jessie said, concluding her story, "is that Chang had never really paid that much attention to Lady. We'd had them both for about five years, and sometimes Lady acted like she wanted to play, but Chang just remained aloof, like messing around with a wiener dog was beneath his dignity. Either Chang loved Lady more than he put on—or else he just hated pit bulls something fierce."

*T*he unspeakable acts of terrorism which took place against the twin towers of the World Trade Center in New York on September 11, 2001, will forever leave a scar on all of our lives. Most of us could never imagine in our wildest dreams such a plot of destruction that would take the lives of thousands of innocent people and leave many more thousands mourning the loss of their loved ones.

For us, the horrible event became intensely personal as we witnessed through the graphic medium of television the collapse of the first and second towers of the World

Trade Center, for our son Steve and his wife, Melissa, lived but blocks away from the towers. We eventually learned of our own miracle: Steve and Melissa were safe, but they had come within minutes of being at the Trade Center's famous restaurant, Windows on the World, for a scheduled meeting.

The many days, weeks, and months that followed the search-and-rescue efforts in the terrorist-ravished area continued to provide both ghastly and frightful images, as well as those of miraculous recoveries through the tireless efforts of the firemen and volunteer workers in the midst of the devastation and danger to their own lives. Although the media focused on the human lives involved, many other stories of pain and joy also came to light regarding the lives lost or saved of cherished and beloved pets. The following story is such an account.

"This is the first good story I've heard" said Suffolk County Society for Prevention of Cruelty to Animals Chief, Roy Gross, referring to the rescue of Precious, a beautiful pedigree white Persian cat from the debris of a severely damaged apartment building across from the World Trade Center. Precious was taken to the nearby Suffolk County SPCA van that was on the scene to treat search-and-rescue dogs for exposure and exhaustion.

Precious was dubbed "the miracle kitty" by the *New York Post* when its October 18, 2001, issue reported the amazing story of survival. Precious had multiple injuries and was so dirty, confused, dazed, and dehydrated that SPCA Chief Gross could hardly believe the stamina of the cat to survive eighteen days without food. As he and others later determined, Precious had most likely drunk from pools of contaminated water atop the wreckage to stay alive.

D. J. Kerr, owner of Precious, told the *New York Post* that the survival of her beloved cat was unbelievable. "It's a miracle," she said. "I can't believe she is alive." D. J. and her husband, Steve, had already experienced a miracle in that they were out of town when the tragedy occurred. After so many days and nights has passed, they had given up on their Precious, assuming she was dead.

The Kerrs had been out of town for a long weekend, and they had left Precious secure and safe in the care of a house sitter who was to have checked in on Precious at 10:00 on the morning of the terrorist attack. Sadly and needless to say, the house sitter didn't make it, since the atrocity took place shortly after 9:00 A.M. Judging by the extent of the damage done to their apartment, the Kerrs didn't see how their cat would have survived.

Their loft apartment in the seven-story apartment building on Liberty Street was in the direct blast and backlash from the Trade Center, and all the windows were completely blown out. Fire and water had ravaged the interior of the apartment and filled it with bits of metal, dust, and smoke. The Kerrs were told it would be a minimum of nine months before they could return to inhabit their once comfortable home.

Incredibly, emergency workers at Ground Zero told of someone hearing the meowing cries of a cat atop the apartment building on Saturday night, September 29, *eighteen* full days after the attack. Responding to the possibility of saving the cat, the workers sent a recovery dog to do some fancy maneuvering and to locate the cat, which was most likely injured, and bring it to safety.

The recovery dog did its duty, and soon a tattered cat was taken to the SPCA van. With burnt paws from the intense heat, her eyes badly damaged from flying debris, glass, metal, and dust, and having lost two of its nine pounds, Precious would recover from the harrowing ordeal and be reunited with her surprised and elated owners.

The Kerrs welcomed Precious home with her favorite treat, turkey. She ate turkey and drank water like she was making up for the eighteen days she managed to stay alive with no food or good water. Precious purred throughout her entire homecoming meal, contented to be with her loving owners. Precious, as a Persian, a breed said by some to be somewhat snooty and elitist, probably even overlooked the fact that it was a *dog* that carried her in its slobbery mouth from the top of the ruins to safety!

A heartfelt tribute and a great debt of gratitude go out to the wonderful people of the American Society for the Prevention of Cruelty to Animals. Their rescues of pets from homes, apartments, and buildings damaged or lost to the Manhattan terror attack goes far beyond the miracle of saving Precious the cat. Over 150 pets were rescued and treated for multiple injuries—primarily shock, dehydration, breathing problems, and burns. The ASPCA also set up a pet rescue hotline for concerned and worried owners, as well as a counseling service for those individuals who weren't as fortunate as the Kerrs were. The pets rescued included rabbits, dogs, reptiles, and of course *cats!*

About the Authors

BRAD STEIGER is the author/coauthor of over 100 books with millions of copies in print, covering such diverse subject matter as biographies, true crime, and the paranormal. A former high school/college English and creative writing teacher, Brad's early success as a published author launched him into writing full-time and professionally since 1963. In 1978, Brad's book *Valentino* was made into a motion picture by British film director Ken Russell, starring Rudolf Nureyev. Later that same year, Brad co-scripted the documentary film *Unknown Powers*, winner of the Film Advisory Board's *Award of Excellence* in 1978. Brad is considered one of the leading experts in the field of metaphysics and the paranormal. Among Brad's honors: Metaphysical Writer of the Year, Hypnosis Hall of Fame, and Lifetime Achievement Award.

SHERRY HANSEN STEIGER, author/coauthor of twenty-nine books has an extensive background as diverse and varied as a writer, creative director for national advertising agencies, magazine editor, producer, model, and actress to studying the healing arts and theology—traditional and alternative. A former teacher, counselor, and an ordained minister, Sherry co-created and produced the highly acclaimed *Celebrate Life* multi-awareness program, performed around the country for colleges, businesses, and churches, from the mid-1960s on, and established one of the earliest nonprofit schools with new approaches to healing body, mind, and spirit—*The Butterfly Center for Transformation*. Among her honors: International Woman of the Year from Cambridge, and Five Hundred Leaders of Influence-Twentieth Century Achievement Award, on permanent display in the U.S. Library of Congress.

Brad and Sherry have been featured in twenty-two episodes of the popular television series, *Could It Be a Miracle*. Together their television appearances and specials include: *The Joan Rivers Show, Entertainment Tonight,* HBO, *Inside Edition, Hard Copy, Hollywood Insider,* USA network, the Arts and Entertainment (A & E) Channel, and the Learning Channel, among others. The Steigers write a monthly angel column for Beliefnet.com and are frequent guests on international radio talk shows.

Visit the Steigers' Web site at *www.bradandsherry.com.*

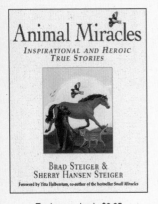